Stimuli-Responsive
Polymer Systems

Stimuli-Responsive Polymer Systems—Recent Manufacturing Techniques and Applications

Special Issue Editors

Akif Kaynak
Ali Zolfagharian

MDPI • Basel • Beijing • Wuhan • Barcelona • Belgrade

Special Issue Editors

Akif Kaynak
Deakin University, Faculty of Science, Engineering,
Australia

Ali Zolfagharian
Deakin University, Faculty of Science, Engineering,
Australia

Editorial Office
MDPI
St. Alban-Anlage 66
4052 Basel, Switzerland

This is a reprint of articles from the Special Issue published online in the open access journal *Materials* (ISSN 1996-1944) from 2018 to 2019 (available at: https://www.mdpi.com/journal/materials/special_issues/srpsrmta).

For citation purposes, cite each article independently as indicated on the article page online and as indicated below:

LastName, A.A.; LastName, B.B.; LastName, C.C. Article Title. *Journal Name* **Year**, *Article Number*, Page Range.

ISBN 978-3-03921-483-9 (Pbk)
ISBN 978-3-03921-484-6 (PDF)

© 2019 by the authors. Articles in this book are Open Access and distributed under the Creative Commons Attribution (CC BY) license, which allows users to download, copy and build upon published articles, as long as the author and publisher are properly credited, which ensures maximum dissemination and a wider impact of our publications.

The book as a whole is distributed by MDPI under the terms and conditions of the Creative Commons license CC BY-NC-ND.

Contents

About the Special Issue Editors vii

Akif Kaynak and Ali Zolfagharian
Stimuli-Responsive Polymer Systems—Recent Manufacturing Techniques and Applications
Reprinted from: *materials* **2019**, *12*, 2380, doi:10.3390/educsci12152380 1

Qiusheng Wang, Guocong Han, Shuqin Yan and Qiang Zhang
3D Printing of Silk Fibroin for Biomedical Applications
Reprinted from: *materials* **2019**, *12*, 504, doi:10.3390/educsci12030504 3

Shi-kai Hu, Si Chen, Xiu-ying Zhao, Ming-ming Guo and Li-qun Zhang
The Shape-Memory Effect of Hindered Phenol (AO-80)/Acrylic Rubber (ACM) Composites with Tunable Transition Temperature
Reprinted from: *materials* **2018**, *11*, 2461, doi:10.3390/educsci11122461 22

Ali Zolfagharian, Akif Kaynak, Sui Yang Khoo, Jun Zhang, Saeid Nahavandi and Abbas Kouzani
Control-Oriented Modelling of a 3D-Printed Soft Actuator
Reprinted from: *materials* **2019**, *12*, 71, doi:10.3390/educsci12010071 35

Sergio Calixto, Valeria Piazza and Virginia Francisca Marañon-Ruiz
Stimuli-Responsive Systems in Optical Humidity-Detection Devices
Reprinted from: *materials* **2019**, *12*, 327, doi:10.3390/educsci12020327 48

Mahdi Bodaghi, Reza Noroozi, Ali Zolfagharian, Mohamad Fotouhi and Saeed Norouzi
4D Printing Self-Morphing Structures
Reprinted from: *materials* **2019**, *12*, 1353, doi:10.3390/educsci12081353 66

About the Special Issue Editors

Akif Kaynak is a leading researcher in stimuli-responsive polymers with soft actuator applications within the School of Engineering, Deakin University, Australia.

Ali Zolfagharian is a Mechanical Engineering lecturer with expertise on 3D/4D printing of soft robots and soft actuators within the School of Engineering, Deakin University, Australia.

Editorial

Stimuli-Responsive Polymer Systems—Recent Manufacturing Techniques and Applications

Akif Kaynak * and Ali Zolfagharian *

School of Engineering, Deakin University, Geelong, Victoria 3216, Australia
* Correspondence: Akif.kaynak@deakin.edu.au (A.K.); a.zolfagharian@deakin.edu.au (A.Z.)

Received: 23 July 2019; Accepted: 25 July 2019; Published: 26 July 2019

Keywords: stimuli-responsive polymer; soft robotic actuators; 3D printing; 4D printing

---

Stimuli-responsive polymer systems can be defined as functional materials that show physical or chemical property changes in response to external stimuli, such as temperature, radiation, chemical agents, pH, mechanical stress, and electric and magnetic fields. Recent developments in manufacturing techniques facilitated the production of different types of stimuli-responsive polymer systems, such as micro- and nanoscale structures with potential applications in soft sensors and actuators, smart textiles, soft robots, and artificial muscles. This special issue presents one review and four scientific report articles. In the review article [1], Wang's group from Key Laboratory of Textile Fiber and Product in Wuhan Textile University evaluates the requirements and characteristics of silk fibroin (SF) as a three-dimensional (3D) printing bioink in biomedical applications. The current challenges of cell-loading SF-based bioinks are comprehensively viewed from their physical properties, chemical components, and bioactivities. The article provides an overview of the programmable and multiple processes involved, including suggestions for further improvement of silk-based biomaterials fabrication by 3D printing. Hu's group from Beijing University of Chemical Technology presents a paper on the preparation and processing of novel polymer materials to develop a shape memory rubber composite with a tailorable transition temperature and excellent shape recovery and fixity [2]. The proposed approach of adjusting the transition temperature of responsive rubber composites enables new design possibilities in stimuli-responsive polymer systems.

Zolfagharian's group from the School of Engineering in Deakin University demonstrates the applications of stimuli-responsive polymers, particularly polyelectrolyte hydrogels, in a soft robotic actuator, which is developed by 3D printing technology [3]. Due to parametric uncertainties of such actuators, which originate from both the custom-design nature of 3D printing and the time variant characteristics of polyelectrolyte actuators, a sophisticated model to estimate their behavior is developed. A practical system identification-based modeling approach for the deflection of the 3D-printed soft actuators incorporating Takagi–Sugeno (T–S) fuzzy sets is proposed and successfully tested in response to a broad range of input voltage variations. With some modifications in the electromechanical aspects of the model, the proposed modeling method can be used with other 3D-printed stimuli-responsive polymer systems. In the fourth article, Calixto's group presents the application of stimuli-responsive materials in electronic devices to measure Relative Humidity (RH) [4]. Gelatin and interpenetrated polymers are utilized to develop an RH detector with a spark-free optical method. The water vapor is used as a stimulus to change film thickness and its refractive index. To detect the change of these two parameters, an optical method based on diffraction gratings is employed.

The special issue closes with the application of stimuli-responsive polymer systems in four-dimensional (4D) printing. Bodaghi's group from the Department of Engineering in Nottingham Trent University presents the emergence of 4D-printed self-morphing structures manufactured by stimuli-responsive and shape memory polymers [5]. The article discusses harnessing complex structures with self-bending/morphing/rolling features fabricated by 4D printing technology, and replicate their thermo-mechanical behaviors using a simple computational tool. Fused deposition modeling (FDM) is implemented to fabricate adaptive composite structures with performance-driven functionality built directly into materials. The effects of printing speed on the self-bending/morphing characteristics are investigated in detail. Thermo-mechanical behaviors of the 4D-printed structures are simulated by introducing a straightforward method into the commercial finite element (FE) software package of Abaqus, which is much simpler than writing a user-defined material subroutine or an in-house FE code. Finally, the developed digital tool is implemented to engineer several practical self-morphing/rolling structures.

Funding: This research received no external funding.

Acknowledgments: As the Guest Editors we would like to thank all the authors who submitted papers to this Special Issue. All the papers submitted were peer-reviewed by experts in the field whose comments helped improve the quality of the edition. We would also like to thank the Editorial Board of Materials for their assistance in managing this Special Issue.

Conflicts of Interest: The authors declare no conflict of interest.

References

1. Wang, Q.; Han, G.; Yan, S.; Zhang, Q. 3D Printing of Silk Fibroin for Biomedical Applications. *Materials* **2019**, *12*, 504. [CrossRef] [PubMed]
2. Hu, S.-K.; Chen, S.; Zhao, X.-Y.; Guo, M.-M.; Zhang, L.-Q. The Shape-Memory Effect of Hindered Phenol (AO-80)/Acrylic Rubber (ACM) Composites with Tunable Transition Temperature. *Materials* **2018**, *11*, 2461. [CrossRef] [PubMed]
3. Zolfagharian, A.; Kaynak, A.; Yang Khoo, S.; Zhang, J.; Nahavandi, S.; Kouzani, A. Control-oriented modelling of a 3D-printed soft actuator. *Materials* **2019**, *12*, 71. [CrossRef] [PubMed]
4. Calixto, S.; Piazza, V.; Marañon-Ruiz, V.F. Stimuli-Responsive Systems in Optical Humidity-Detection Devices. *Materials* **2019**, *12*, 327. [CrossRef] [PubMed]
5. Bodaghi, M.; Noroozi, R.; Zolfagharian, A.; Fotouhi, M.; Norouzi, S. 4D Printing Self-Morphing Structures. *Materials* **2019**, *12*, 1353. [CrossRef] [PubMed]

© 2019 by the authors. Licensee MDPI, Basel, Switzerland. This article is an open access article distributed under the terms and conditions of the Creative Commons Attribution (CC BY) license (http://creativecommons.org/licenses/by/4.0/).

Review

3D Printing of Silk Fibroin for Biomedical Applications

Qiusheng Wang, Guocong Han, Shuqin Yan * and Qiang Zhang *

Key Laboratory of Textile Fiber & Product (Ministry of Education), School of Textile Science and Engineering, Wuhan Textile University, Wuhan 430200, China; qiusheng-wang@hotmail.com (Q.W.); han792464210@hotmail.com (G.H.)

* Correspondence: ysq_zq@163.com (S.Y.); qiang.zhang@wtu.edu.cn (Q.Z.)

Received: 1 January 2019; Accepted: 2 February 2019; Published: 6 February 2019

Abstract: Three-dimensional (3D) printing is regarded as a critical technological-evolution in material engineering, especially for customized biomedicine. However, a big challenge that hinders the 3D printing technique applied in biomedical field is applicable bioink. Silk fibroin (SF) is used as a biomaterial for decades due to its remarkable high machinability and good biocompatibility and biodegradability, which provides a possible alternate of bioink for 3D printing. In this review, we summarize the requirements, characteristics and processabilities of SF bioink, in particular, focusing on the printing possibilities and capabilities of bioink. Further, the current achievements of cell-loading SF based bioinks were comprehensively viewed from their physical properties, chemical components, and bioactivities as well. Finally, the emerging issues and prospects of SF based bioink for 3D printing are given. This review provides a reference for the programmable and multiple processes and the further improvement of silk-based biomaterials fabrication by 3D printing.

Keywords: silk fibroin; 3D printing; bioink; properties; biomedical applications

---

1. Introduction

In recent years, three-dimensional (3D) printing is a promising strategy to the biomedical field and it is regarded as a future alternative to current clinical treatments. Not only that it can alleviate the artificial organ or tissue shortage crisis, but it can also design and produce complex and precise microstructures according to reconstruction of tissue engineering requirements [1–3]. More importantly, a series of advanced 3D printing techniques have been approved to achieve structural and functional consistency with model design, which means that competitive manufacturing technology is ready for tissue repair and transplantation [4–6]. Bioink as a core of the 3D printing is the key to success for 3D printing products. Specifically, bioinks loading cells, growth factors, and cues for bio-applications are still in the early stage in 3D printing. Therefore, it is an urgent need to seek an appropriate material as bioink for 3D printing.

Bioinks are cell-encapsulating biomaterials that are used in 3D printing process and they must be friendly to both printing process and 3D cell culture [7]. However, most of biomaterials are insufficient in meeting requirements of ideal bioink, so that choosing a suitable biomaterials as bioink plays an significant role in rebuilding the similar function of native tissue following the principle of tissue engineering [8]. In the field of tissue engineering, the three strategies that were used to replace or repair native tissue: using cells, cytokines, or cell substitutes only; using biocompatible biomaterials only to induce tissue regeneration; combination of using cells, cytokines, and biomaterials [9]. Thus, including non-toxic, cytocompatibility, bioactivity, free-standing, and applicable mechanical properties, and cell-loading and encapsulation ability in the physiological conditions, are the pre-requirements and properties of the biomaterial as a bioink. Additionally, when considering the sustainable process of printing, the printability of bioink depends on several controllable parameters, including the viscosity of solution, the ability of crosslinked, and surface tension of the bioink. If the viscosity of the bioink formulation is higher, a larger pressure is needed for the extrusion of bioink from the small nozzle, or causing the nozzle to be blocked and cell death [10,11]. On the other hand, the crosslink mechanism and surface tension are critical to cell's activity, aggregation, and viability. From the perspective of the biomedical field, time-consuming is a vital factor and can never be ignored, especially in cell-based printing. It usually results a decrease in cell viability for preparation of scaffolds with large and complex structures by 3D printing [12]. The cell-based and cell-free approaches are two categories of bioink used in 3D printing, thus the cell carrier or tissue substitute should keep a balance between self-digestion and tissue regeneration [13,14]. A tunable biodegradability should be taken into consideration, so that the rate of tissue regeneration can be matched. Finally, easy manufacturing or processing that are affordable and readily available are encouraging and welcoming features for selecting suitable biomaterials as bioink formulation [15].

Following the rules of ideal bioink, several cases have demonstrated that hydrogels with a high content of water and shape plasticity are attractive candidates as bioinks [16–18]. Based on the features, including bio-instructive, cell encapsulation, and a 3D microenvironment, many hydrogels have been developed from naturally derived polymers, such as gelatin, fibrin, collagen, chitosan, alginate, and hyaluronic acid (HA) [19–23]. The gelation mechanism by chemical crosslinking (for gelatin and hyaluronic acid) and ionic (for chitosan and alginate) are not suitable for the bioactivity of cell-loading bioink, and the inappropriate degradation rate (for fibrin and collagen) also shows an unfavorable servicing. Previously, a series of Silk fibroin (SF) products gained much attention for application and they were studied as a protein polymer for biomedical applications, for instance, in the enzyme immobilization matrix [24], wound dressing [25], vascular prosthesis [26], and artificial grafts [27], due to its similar components to the extracellular matrix (ECM), low-cost, tunable mechanical properties, controllable degradation, and good biocompatibility [28,29]. The timeline of the development of SF based bioink in 3D printing technology [30–36] over the past 30 years has witnessed great research and application value for the customized biomedical filed (Figure 1). These results encouraged further exploration of the SF based biomaterials via 3D printing.

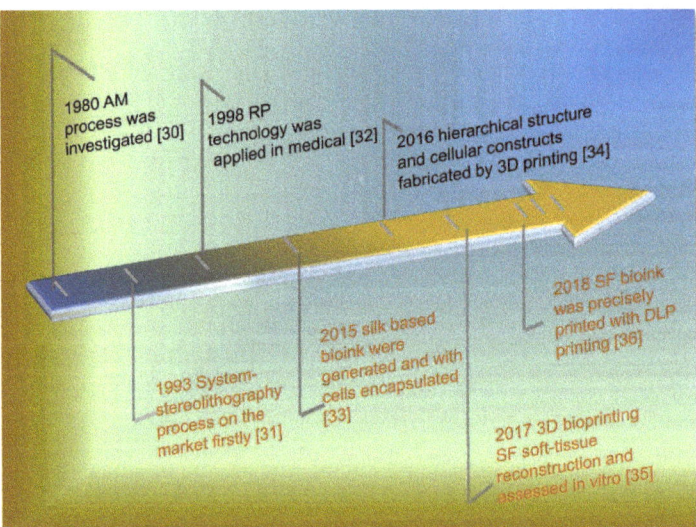

Figure 1. Timeline of the development of Silk fibroin (SF) based bioink in three-dimensional (3D) printing technology [30–36]. Additive manufacturing (AM); Rapid prototyping (RP); and, Digital light processing (DLP).

In this review, we firstly discuss the evolution toward 3D printing derived from SF (*Bombyx mori* silkworm) bioink, mainly focusing on the improvement and design of SF bioink to match the requirements of ideal bioink. Subsequently, we summarize the advanced progress in biomedical applications that are based on 3D printing of SF bioink in vitro. Finally, we outlook the broader challenges and directions for the future development of SF bioink for functional materials designs and engineering via 3D printing.

2. Silk Fibroin Bioink

2.1. Processing of SF Bioink

Native *B. mori* silk is composed of silk fibroin protein coated with sericin protein, and sericin is a group of soluble glycoproteins that are expressed in the middle silk gland of *B. mori* silkworms [16]. By degumming, the sericin is removed, the SF fibers could be dissolved and purified into an aqueous solution through dialyzing in deionized water [37]. Based on aqueous solution system, the SF can be further processed into different types of materials in films, particles, fibers, and sponges, also including hydrogel. However, there is a barrier hindering 3D printing fabrication in SF bioink that is caused by low concentration and viscosity. Increasing its concentration and adding other high viscosity additives are perhaps useful strategies in improving its printing processability and biofunction ability.

To obtain high concentration SF solution, as shown in Figure 2, there are two approaches that are employed. One way is based on the SF purification protocol that is modified with some additional procedures. Specifically, SF solution is concentrated with a dialysis bag (Molecular Weight Cut Off (MWCO) ≈ 3000 Da) in polyethylene glycol (PEG, Molecular Weight (MW) ≥ 20000 Da) solution, or regenerated SF materials are re-dissolved in organic solvents (1,1,1,3,3,3-hexafluoro-2-propanol (HFIP), Formic acid, etc.) to increase the concentration to meet the requirements of rheology of bioink [17,18]. However, the bioactivity of silk proteins will be inevitably weakened by these complexing processes. Recently, adapting new dissolving systems for another effective way, the Ca^{2+}-formic acid binary solvent system and HFIP are studied as dissolving solvent directly for silk fibers to produce high concentration SF solution [38,39], which is easier for yielding over 20 wt.%. These unfriendly solvents will continue cutting the SF molecules chains in a further process, resulting in low SF molecular weight and viscosity. What is more, the unfriendly solvent residues have a detrimental effect on cell viability and encapsulating in 3D printing, which limited the applications of these solvents in 3D printing. As a second strategy, it is convenient and highly efficient to enhance the free-standing and viscosity of SF based bioink by blending other high viscosity biomaterials. Based on the principle of similar compatible, gelatin, chitosan, alginate, and HA are mixed with SF solution to prepare SF based bioink [33,36,40]. This strategy is more successful than other approaches in improving the SF solution with a high concentration and plastic ability for 3D printing.

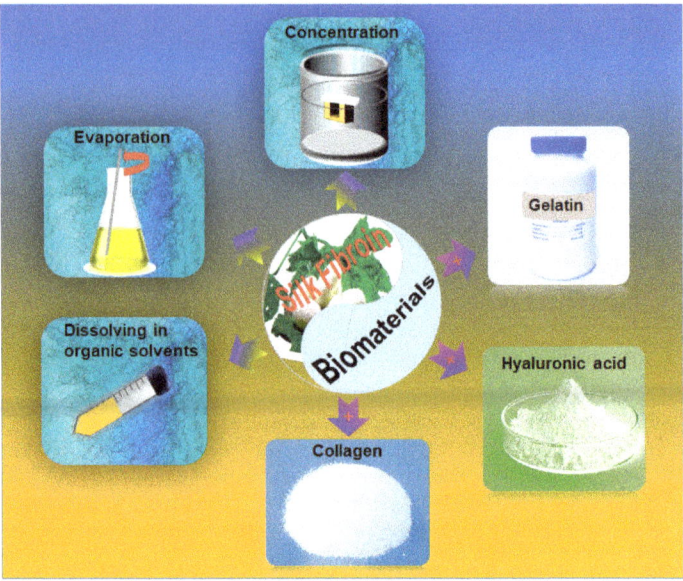

Figure 2. Schematic of methods to optimizing the rheology of SF bioink. SF is a biomaterial with impressive biocompatibility and mechanical properties. As a bioink, its rheology should be adjusted in aqueous system by different strategies. The gradient arrow without "+" indicated that their rheology could be regulated by concentration, evaporation, and dissolving in organic solvents; the arrow with "+" shows that SF solutions were combined with other biopolymers, such as collagen, hyaluronic acid, and gelatin, respectively, to enhance their rheology.

2.2. SF Bioink Design

Nowadays, although synthetic polymers broaden the diversity of materials, their low cell viability and non-biocompatible degradation products hinder making a further step as bioinks. Natural materials, like cellulose, HA, and collagen, are friendly to cell growth and development as SF materials, while the slow gelation rate or inappropriate mechanical properties always mismatch with rapid additive manufacturing technology [41,42]. Encouraging by the easy processing and abundant source, SF, as a bioink, motivated more researchers to explore their wide range of applications.

By contrasting with the characteristics of SF and polymers that are mentioned above (Table 1), single-SF is probable to yield into bioink for 3D printing in aqueous system. According to the LiBr-dissolving protocols, SF bioink is treated to optimize its rheological ability via the purification and concentration process by slowly stirring and low temperature evaporation, and their mechanical properties and degradation could be controlled by the regulation of β-sheet content, degree of crosslinking, and morphological structures [43,44]. Nature silks have showed a lot of features, such as outstanding strength and toughness, controllable degradation, and high cell viability (Figure 3). The regenerative SF materials usually resulted in the deterioration of mechanical properties, which could be reinforced by inducing conformation transition. Specifically, several approaches are employed for transformation random coil or helical conformation into β-sheet structure to induce the SF insoluble, such as alcohol solution treatment [45], soft-freezing treatment [46], shear force inducing [47], salts addition and crosslinker [48], and pH value adjustment [49]. These approaches may be used to enhance the free-standing of SF 3D printing scaffolds and regulate their biodegradation in vitro and in vivo. These characteristics also indicated that the printability and mechanism of SF bioink could be controlled to meet different printing purposes.

Table 1. Comparative analysis of silk versus other pure polymeric bioinks.

Materials	Advantages	Disadvantages	Crosslinking Methods
Silkworm silk	i. Ease of structure modification [37] ii. Controlled degradation iii. High cellular viability iv. Diversity of methods for crosslink or sol-to-gel [50] v. Outstanding strength and toughness vi. Embedded hydration properties [28] vii. Abundant sources	i. Rheology need to be optimized as bioink [51] ii. Low viscosity [52] iii. Hard to printing individually	i. Enzymatically ii. Temperature iii. pH value changes iv. Sonication v. Salting leaching vi. Photo-crosslink
Alginate	i. Ease of crosslinking ii. Stability of constructs iii. Biocompatible, facilitates cell entrapment iv. Ease of processability [53]	i. Fast degradation in vitro, need additional dopants ii. Low cell attachment and protein adsorption iii. Lack of adequate mechanical properties	i. Ionic (Ca^{2+})
Agarose	i. Non/low-toxic ii. Biological properties can be improved with other hydrogel easily iii. Suitable mechanical properties for cartilage tissue repairing [54]	i. Non-degradable ii. Not suitable for inject printing with high viscosity iii. Low cell adhesion and spreading	i. Low temperature
Collagen	i. Easy degradation ii. Facilitate cell adhesion and cell attachment iii. Easy to modify with other polymers iv. Need to improve its mechanical and biological properties with other polymers [41]	i. Time-consuming for gelation ii. Complex process to purification iii. Low mechanical properties iv. Biorisk	i. pH ii. Temperature iii. Vitamin Riboflavin iv. Tannic acid [55]
Fibrin	i. Excellent biocompatibility and biodegradation [56] ii. Rapid gelation iii. Easily purified from blood providing autologous source iv. Superior elasticity	i. Weak mechanical properties ii. Severe immunogenic responses iii. So fast for its degradation	i. Enzymatic treatment

Table 1. Cont.

Materials	Advantages	Disadvantages	Crosslinking Methods
Cellulose	i. High mechanical properties ii. Helpful for improving cells viability [57] iii. Excellent shape fidelity [58]	i. Environment sensitive ii. Non-biodegradation in vivo iii. Purification	i. Ca^{2+}
Hyaluronic acid	i. Fast gelation ii. Controllable mechanics, architecture, and degradation iii. Supports cell adhesion, migration, proliferation [59]	i. Weak mechanical properties ii. Need chemical modification to regulate the rheology.	i. Photo-crosslink
Hydroxyapatite	i. Keep good shape fidelity and produce porous [60]	i. Slow degradation rates [61] ii. Low bioactivity	i. Methanol

When considering that function of biomaterials in the reconstruction of neo-tissue by providing a stable and biocompatible microenvironment for cells proliferation and differentiation in tissue engineering [62], the bioink should be designed intensively. SF is one of the most studied and industrially used types of fibrous proteins in biomedical applications. Several attempts have been made in biomedical with 3D printing technology. However, some aspects of SF bioink should be addressed based on previous cases. Specifically, from the point of a physic-chemical view, the printability of bioink should take care of some parameters, including rheology, swelling ratio, and surface tension [14]. First, the excellent rheology is the basic requirement for bioink that was extruded from the nozzle, as the higher extruded-forces would harm cell viability [63]. The proper swelling ratio is beneficial to the formation of certain two-dimensional (2D) morphological structure after the bioink extruded, which have a role in improving resolution and free-standing of printing products. Third, more attention should be paid to surface tension, which exists between the compounds that are present in the liquid. It plays a big role in building a 3D structure for cell attachment distribution and development [64]. The surface tension should be self-adjustable so as to meet the changes that the surface tension imposes on the liquid-gas interface [14]. Moreover, from a bio-fabrication point of view, the excellent cell-encapsulating or growth factors-loading abilities are significant for cell proliferation and adhesion. Hence, the SF bioink based on aqueous system or cell culture medium system should put more efforts into retaining them in future studies.

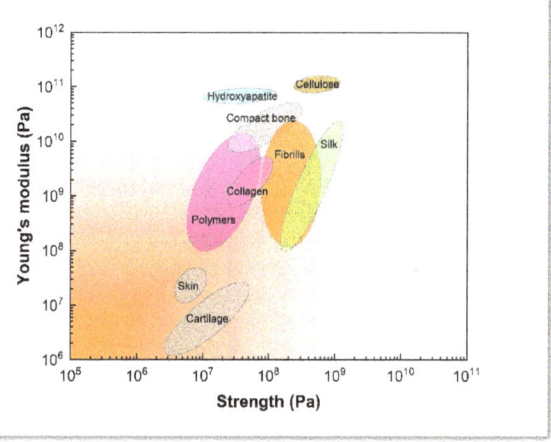

Figure 3. Comparison of the specific values of strength and stiffness of SF materials with natural and synthetic materials. Reproduced with permission from [65]. Copyright 2011, Nature.

Regarding the bio-inspiration of silkworm spinning, the process of silk cocoons formation is a typical procedure of architecture by the 3D printing technique. There is no doubt that silk protein solution is an ideal and attractive choice for bioink [40]. Because of the existence of sericin, silks are easy to spin and build into the silk cocoons approach to 3D printing by silkworm. The natural behavior of silkworms highlights to us that single component SF is insufficient for 3D printing. Blending and hybrid bioink should be considered in improving in the aspects of printability, especially for rheology and viscosity [51,66]. Wet spinning or microfluidic spinning cases demonstrated that the two factors for rheology and viscosity of fluid included deformation energy stored (G′) and dissipated energy (G″) [67,68]. As shown in Figure 4, the SF G′ always exceeded G″ at high frequencies and vice versa at low frequencies, which means that it is conductive as viscoelastic liquid, and these characteristics determine the rheology of multicomponent bioink [69].

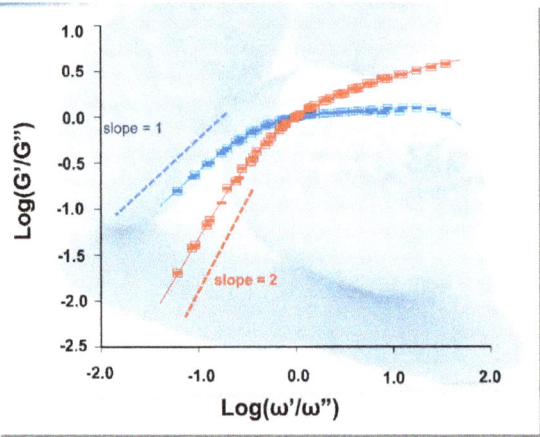

Figure 4. The relationship between loss factor and frequencies of SF. Reproduced with permission from [69]. Copyright 2016, American Chemical Society.

The basic physical characteristics of SF bioink should not only be addressed, but some chemical characteristics are also helpful in optimizing its printing abilities, especially in self-assembly [70], chemical decorative [71], and conformation transition induction. On one hand, once the amino acid sequence of SF self-assembled into an antiparallel β-sheet structure by intra- and inter-molecular hydrogen bonds [72], which would contribute to robust mechanical properties. On the other hand, the presence of several reactive amino acids in SF allow for easily accessible chemical modification strategies, including coupling reactions [73], amino acid modification [74], and grafting reactions [75]. Based on chemically modifiable of SF, the recently report showed that SF could be modified with methacrylate groups directly for light polymerization, which would be beneficial to improve its printability [36]. These strategies are utilized to tailor the protein for a desired function or form [76]. Based on physical and chemical characteristics of SF solution, SF bioink shows a strong vitality for 3D printing when it combined with other biomaterials via optimizing the basic parameters of the bioink, such as printability, mechanical properties, shape fidelity, and cell viability [77] (Table 2). Raw material screening and formula optimization usually are the initial and essential steps in multicomponent bioink. As mentioned before, the combination of SF with polysaccharide bioink is an effective approach to adjust rheological properties, such as chitosan, alginate, and HA. The gelation rate and printability can be improved significantly with alginate being applied as an additive component [78]. Gelatin as another great candidate for modulating SF based bioink properties gains much attention due to its similarity to human ECM and with a gentle gel environment, and its rapid degradation rate and weak mechanical properties are supported by the incorporation of SF [79]. Therefore, it will promote 3D printing technology to develop a SF based-multicomponent bioink to overcome the shortages of single bioink.

Besides the basic physical and chemical characteristics of SF bioink, the biological performance is another essential indicator that can never be ignored in bioinks. Over past decades, numerous studies witnessed and proved the excellent biological properties of SF, and properly degummed and sterilized silk manufactures demonstrated biocompatibility and bio-viability that were as good as commercial products of polylactic acid and collagen [80]. The United States (U.S.) Food and Drug Administration approve of these products. SF bioink performances are described as followed in: (1) huge cell-loading printability for precisely control SF ink deposition [81], which has advantages in overcoming the uncontrollable cell dynamics [82]; the mismatch of printing process parameters [83]; (2) the good encapsulation ability for cells, drugs, and bioactive molecules [84,85]; and, (3) excellent viability for different cells and cell lines for proliferation and differentiation [86].

Table 2. The properties of bioink formulated by multicomponent materials based on SF.

Bioink Formulation	Crosslink Method(gelation)	Cell Types & Density & Viability	Advantages (A) and Disadvantages (D)	Applications	Printing Method	Ref.
SF-Gelatin	Enzymatic/sonication	hTMSCs; BMSC 2.5×10^6 mL^{-1}; 2×10^5 86% (30 days); enriched (21 days);	A: Enhances cell adhesion Good mechanical	Artificial Implant/Cartilage tissue engineering	Inject printing	[50,79,87]
SF-Collagen	Ethanol	BMSCs 2×10^7 cells 4×10^2 cell (13 days);	A: Comprehensive physical properties; support cell growth	Knee cartilage; Tissue engineering	Extrude printing	[88]
SF-Chitosan	hexamethylene diisocyanate/ chlorohydrin/ glutaraldehyde	BMSCs 2×10^7 mL^{-1} 10^2 cells;	A: Produce high porosity with different structures; D: the cross-linking agent have cytotoxic	Tissue engineering Drug release	Extrude printing	[88]
Cartilage acellular matrix (CAM)-SF	Enzyme (EDC-NHS)	rBM-MSCs Seeding efficiency 65% >80%;	D: Poor shape fidelity; low precision of printing	Cartilage tissue engineering	Extrude printing	[89]
SF-Alginate	Horseradish peroxidase (HRP)-H_2O_2	NIH3T3 5×10^5 mL^{-1} begin to decline slowly (42 days);	A: maintain long-term metabolic activity for bioink D: the compatibility of silk and alginate need to be improved.	Vascular tissue engineering	Inject printing	[78]
SF/polyethylene glycol (PEG)	Sonication	hMSCs 2.5×10^6 mL^{-1} 50% (3 weeks);	A: maintain shape for a long time (6weeks); the crosslinker without damage cell viability; with a good mechanical and high shape fidelity	Cartilage tissue engineering	Inject printing	[90]
SF-glycidyl methacrylate	Photo-crosslink	NIH/3T3 1×10^6 mL^{-1} 50% (4 weeks)	A: a gentle crosslink environment and friendly to cells growth; the mechanical properties improved with Sil-MA concentration increased.	Bone tissue engineering	Digital light printing	[36]

However, as shown in Table 2, choosing a suitable 3D printing technology and method is also the key to the success of SF bioink. Firstly, the concentration and rheology of SF should match the requirements of different additive manufacturing technology. For instance, the low viscosities (3.5–12 mPas^{-1}) behave in inkjet printing better than extrude printing (about 600 kPas^{-1}) [91,92], and the large mechanical stress that is applied to extrude the bioink resulting in the reduction of cell survival. Secondly, adapting controllable physical gelation or phase transition strategies so that the toxicity of crosslinkers to cells is reduced and cell encapsulation is enhanced [93]. Take the photo-crosslink as an example; it is an effective method improving the cell viability for the chemically modified SF [36]. Thirdly, the printing resolution of SF bioink is susceptible to printing parameters, such as temperature, printing speed, and SF molecular weight. As for as the resolution of 3D printing, laser-based printing have a high resolution of 1–3 µm, but it has to solve cell-damage that is induced by laser [94]. Extrude printing technology with a low resolution (about 100 µm), which was applied more in the most recent researches [95]. Additionally, a great potential for inkjet printing is attached to more future study for its relative high resolution of 50 µm and high printing speed [96].

3. Evaluation of Cell Viability with SF Based 3D Printing Scaffolds

As a bioink, cell viability is another key point of its success in 3D printing. There are several cases regarding 3D printing creations that are based on SF bioink [35,36,50,97–99]. 3D printing artifacts allowed for cell seeding that is more efficient than that of porous scaffolds derived from freeze-drying and electrospinning techniques [100] (Figure 5a); their precise mimic nature tissue framework could regulate cell phenotypes and neo-tissue reconstruction by stimulating cell differentiation and proliferation. The component of SF also acted a positive influence on the biocompatibility and bioactivity of 3D printing scaffolds via providing a different stiffness and rough morphology [97] (Figure 5b). In fact, the microenvironment and time-consumption of printing objects should match with the cell aggregation and proliferation disciplines. In order to maintain long-term cellular viability, desired cellular distribution and mild mechanical action are necessary during 3D printing. Previous studies show that the cell viability of top layers is better than that of central layers after 14 days, which may be ascribed to a 3D open-porous structure that facilitated, at a certain extent, the diffusion of nutrient and oxygen to the encapsulated cells during their residence. In addition, the layer by layer manufacturing technology is a time-consuming process that is influenced by the cell fate of the cell-loading bioink directly [101] (Figure 5c1,2). Secondly, a friendly or low side effective crosslinking method should be adopted in improving cell viability. For instance, the cell viability shows a trend of significant decrease for SF-alginate bioink with crosslinked tyrosinase. The results showed that the tyrosinase-crosslinking has an unfavorable effect on cells encapsulation in the long term (30 days) [50] (Figure 5d1,2). The physical crosslinking method may be a proper approach for 3D printing fabrication. Thirdly, the equal environment should be adapted by multi-cell to predict whether the printed object acts upon an implant, causing immune rejection in the body or not. As shown in Figure 5, the chondrocytes and human mesenchymal stem cells (hMSCs) were cultured on 3D printed silk-gelatin scaffold, respectively, the cellular dispersion increased significantly both kinds cells, but the cellular aggregate changed toward opposite directions [40] (Figure 5e1,2). Finally, the bioactivity and mechanical performance are insufficient at the initial stage or after implantation for a while, which usually caused a cavity or cyst in defect sites by supporting deficiency. Consequently, these primary results inspired us with courage in understanding the biological mechanisms of cells and the fabrication of biomedical materials.

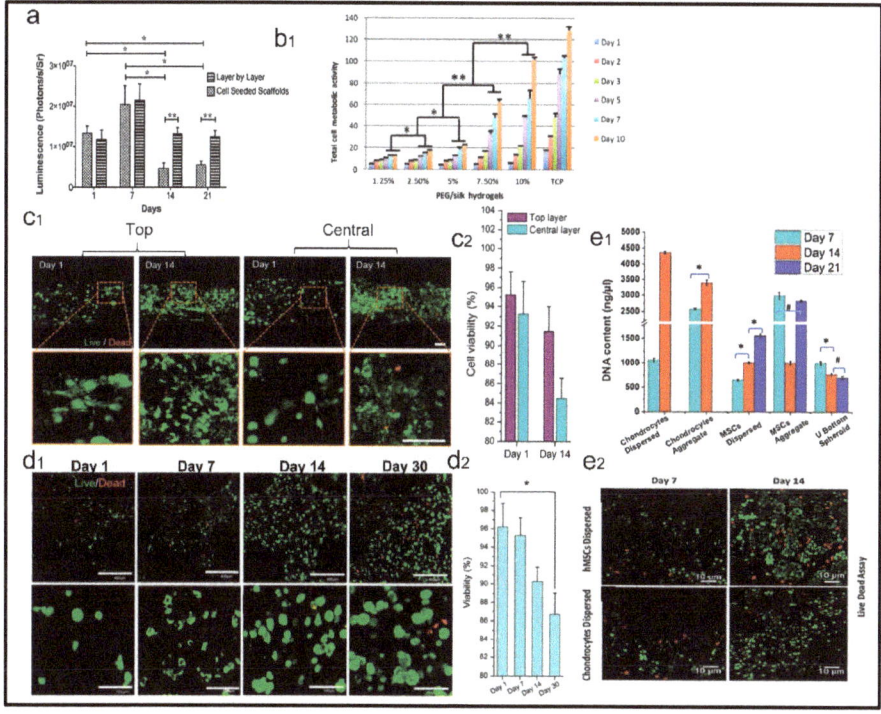

Figure 5. The evaluation of cells with printing bio-model in vitro. (**a**) the cells number assessment of lay-by-layer (LBL) sandwiches and the cell-seeded scaffolds (CSS). Reproduced with permission from [100]. Copyright 2012, Mary Ann Liebert, Inc. (**b**) Total cell metabolic activities with different concentration of SF. Reproduced with permission from [97]. Copyright 2018, Wiley. (**c1,c2**) Cell viability within the decellularized adipose tissue (DAT) constructs in top and layers. Reproduced with permission from [101]. Copyright 2015, Elsevier. (**d1,d2**) Live/Dead images and quantitative analysis of human turbinate mesenchymal stem cells (hTMSCs) that were encapsulated in tyrosinase crosslinked 8SF-15G-T constructs over 30 days. Reproduced with permission from [50]. Copyright 2014, Elsevier. (**e1,e2**) Cell viability and proliferation. Live and dead cell assay of hMSCs and chondrocytes printed with silk-gelatin bioink as dispersed cell at three weeks. Reproduced with permission from [40]. Copyright 2016, American Chemical Society.

4. SF Bioink for Biomedical Applications

4.1. Skin Tissue

With the development of multicomponent bioink and printing technology, a series of biomedical applications have been reported based on the process of 3D printing (Figure 6). The skin is the largest complex organ in the human body and it is composed by three layers (the stratified squamous epithelium, the dermis, and the hypodermis) [102]. Autografts and allografts are two strategies for skin repairing, which is still limited in donors and recipients to some extent. Specifically, the donor suffers from pain, second operation, and scarring, for the recipient, with the exception of scarring, dermal vascularization, and epidermis functionalization, are difficulties facing their subsequent therapy [103]. Recently, a gelatin-sulfonated silk composite scaffold was fabricated by a DIY pneumatic printing system, with the incorporation of growth factors, which presented skin-like tissues and enhanced skin regeneration by printing technology [79]. By the nanoimprint lithography technique, SF film with skin tissue-like nanoscale structures was fabricated to mimic the collagen morphology of natural dermal [29], which is known, as it could alleviate scar formation. The silk-based bioink combined with collagen are also employed to prepare artificial skin grafts, and the network connective of neo-tissue increased alot when compared with scaffolds that are derived from the freeze-drying method [104]. Although SF as bioink to printing artificial skin-tissue is starting out, the available results regarding the histology and immune fluorescence characterization of the 3D printed grafts presented an applicable potential in skin tissue repair.

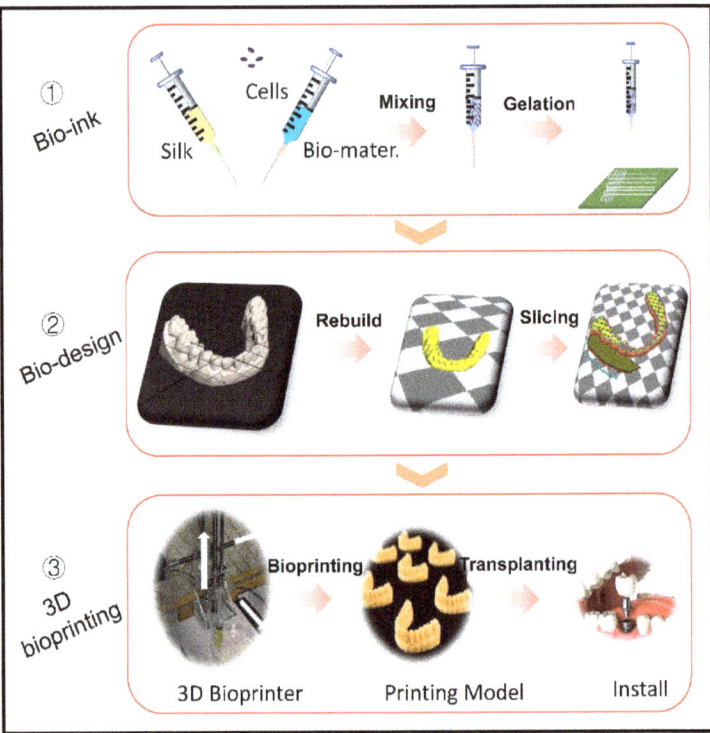

Figure 6. Schematic of 3D printing process. (1) Preparation of SF based bioink for 3D printing (2) Imaging and digital design. (2) 3D printing process and clinical trial. Reproduced with permission from [97]. Copyright 2018, Wiley.

4.2. Cartilage Tissue

Cartilage damage and degeneration are common disease in the aged suffering from osteoarthritis, which has become an urgent need in clinical healthcare [105]. Some challenges still existed in mimicking the fine structures of native cartilage tissue, especially in nano- micro-ordered structures. Fortunately, when comparing to common approaches, the 3D printing fabrication manifested positive practicability [106]. It appeared to be more promising for SF based bioinks with the recent study, though it was not wildly applied in tissue engineering [107]. For example, by integrating SF with gelatin loading growth factors as bioink, it could be optimized in structural and function for cartilage repairing [40,87]. Pure silk bioink with high concentration could be processed by direct-writing technology, which showed that 3D printing is a much more competitive method in resolution, cell viability, and complex tissue formation [108].

4.3. Bone Tissue

Bone tissue engineering usually relies on bone structure, compositions, mechanics, and tissue formation, which makes it crucial in obtaining a fundamental understanding of bone biology [109]. Nevertheless, it has become the focus as to how to keep balance between bioactivity and mechanical properties for printing bone [110]. As for mechanical performance, the bio-ceramics have been used frequently as an important element of bioink, including α-tricalcium phosphate (α-TCP) and hydroxyapatite [61,111,112]. The results showed great potential in bone tissue repair when combined with SF. For instance, polylactic acid/hydroxyapatite/silk ternary bioink, in fabricating bone clip, which demonstrated an equivalent mechanical property, good biocompatibility, and alignment when compared with other types of the bone clip [113]. Another SF/hydroxyapatite scaffold that was fabricated by direct-writing technology and their regular pore size was beneficial regarding cell growth in orientation [29]. A low-temperature printing technology for the collagen/decellularized extracellular/SF scaffold preparation also showed higher cell proliferation and differentiation. When comparing to that of the collagen scaffold, the compressive modulus was highly improved due to the β-sheet formation of SF [114].

4.4. Blood Vessel

Vasculature within the tissues or organs is crucial in transporting oxygen and nutrients and in maintaining tissue functions [115]. Though the quantity demanded is enormous, the thrombogenicity and low patency rate narrowed the clinic utility of artificial blood vessels, especially in repairing small diameter (in 4–6 mm) blood vessels [113]. By convenience of 3D printing, it was greatly encouraged to manufacture blood vessel tissue engineering. SF and glycidyl methacrylate (Sil-MA) as blending bioink was used for building blood vessels in the hydrogel state; the hydrogel showed outstanding mechanical and rheological properties, which provide many possibilities for vessels, brain, and ear with highly complex organ structures [36]. Similarly, the SF incorporated melanin nanoparticles could be as a transparency modifier to adjust poly(ethylene glycol)-tetra acrylate to improve the printing resolution, and these features make it possible to fabricate blood vessels or vacant tubes [116]. In advanced 3D printing technology for fabricating vessels, preference should be given to obtaining enough porosity and mechanical properties and non-thrombosis to combat thrombogenicity at early stage [117]. Therefore, the characteristics of SF bioink should be optimized to satisfy target application and tissue engineering [118].

5. Summary and Prospects

3D printing has become a promising technology and it has gained high and extensive attentions in silk biomaterials. SF, as a natural and ancient protein material, was a great promise candidate for bioink. In this review, we summarized the physical, chemical, and biological characteristics of SF, and deliberated the proceeding methods and contemporary issues of SF as bioink. Although many cases of SF based bioinks have been reported continuously, taking this technology from the bench to the bedside still requires focused efforts on many fronts.

Firstly, because most of the established methods are time-consuming and require a series of chemical reagents, these options can undermine the green and biocompatible features of biomaterials.

Secondly, as a bioink, SF should be designed and processed synthetically, especially in its viscosity, rheology, encapsulation, and biocompatibility. The potential approaches are the concentration of the SF solutions and the incorporation of other biopolymers. With the aim of tissue reconstruction, the various cells and growth factors are highly recommended during 3D printing. A combination of these biopolymers in silk materials can compensate for the limitations of individual components. These have potential to enhance the performance and function of the final materials by 3D printing.

Finally, the homogeneously distribution of the cell before and after printing in bioink is an important parameter to be controlled. From the perspective of manufacturing technology, only by choosing the biomaterials and finding a suitable cell-seeding method can this trouble be resolved. At the same time, optimizing the biodegradation rate of the SF creations to match the speed of neo-tissue regeneration is necessary.

In summary, we established an overall view in understanding the requirements in 3D printing of the SF bioink. The fundamental understanding of this biological ink can accelerate the development of new methods to obtain novel 3D biomaterials and it offers the opportunity for regarding insight protein material designs in biomedical applications.

Author Contributions: Conceptualization, Q.Z. and S.Y.; writing-original draft preparation, Q.W. and G.H.; writing-review and editing, Q.Z. and S.Y. supervised the paper, Q.Z. All authors jointly discussed the results.

Funding: This research was funded by the Nature Science Foundation of Hubei Province, China (2017CFB578, 2018CFB663), the National Nature Science Foundation of China (51303141, 51403163, and 31600774).

Conflicts of Interest: The authors declare no conflict of interest.

References

1. Wu, C.; Wang, B.; Zhang, C.; Wysk, R.A.; Chen, Y.W. Printing: An assessment based on manufacturing readiness levels. *Crit. Rev. Biotechnol.* **2017**, *37*, 333–354. [CrossRef] [PubMed]
2. Mandrycky, C.; Wang, Z.; Kim, K.; Kim, D.H. 3D printing for engineering complex tissues. *Biotechnol. Adv.* **2016**, *34*, 422–434. [CrossRef] [PubMed]
3. Chia, H.N.; Wu, B.M. Recent advances in 3D printing of biomaterials. *J. Biol. Eng.* **2015**, *9*, 4. [CrossRef] [PubMed]
4. Park, S.-H.; Jung, C.S.; Min, B.-H. Advances in three-dimensional printing for hard tissue engineering. *Tissue Eng. Regen. Med.* **2016**, *13*, 622–635. [CrossRef] [PubMed]
5. Guvendiren, M.; Molde, J.; Soares, R.M.; Kohn, J. Designing Biomaterials for 3D Printing. *ACS Biomater. Sci. Eng.* **2016**, *2*, 1679–1693. [CrossRef] [PubMed]
6. Yang, Y.; Song, X.; Li, X.; Chen, Z.; Zhou, C.; Zhou, Q.; Chen, Y. Recent Progress in Biomimetic Additive Manufacturing Technology: From Materials to Functional Structures. *Adv. Mater.* **2018**, e1706539. [CrossRef] [PubMed]
7. DeSimone, E.; Schacht, K.; Pellert, A.; Scheibel, T. Recombinant spider silk-based bioinks. *Biofabrication* **2017**, *9*, 044104. [CrossRef] [PubMed]
8. Ma, P.X.; Eyster, T.W.; Doleyres, Y. Tissue Engineering Biomaterials. In *Encyclopedia of Polymer Science and Technology*; Wiley: Hoboken, NJ, USA, 2016; pp. 1–47. [CrossRef]
9. Ali Khademhosseini, R.L.; Borenstein, J.; Vacanti, J.P. Microscale technologies for tissue engineering and biology. *Proc. Natl. Acad. Sci. USA* **2005**, *103*, 2480–2487. [CrossRef]
10. Gopinathan, J.; Noh, I. Recent trends in bioinks for 3D printing. *Biomater. Res.* **2018**, *22*, 1–15. [CrossRef]
11. Freeman, F.E.; Kelly, D.J. Tuning Alginate Bioink Stiffness and Composition for Controlled Growth Factor Delivery and to Spatially Direct MSC Fate within Bioprinted Tissues. *Sci. Rep.* **2017**, *7*, 17042. [CrossRef]
12. Norotte, C.; Marga, F.S.; Niklason, L.E.; Forgacs, G. Scaffold-free vascular tissue engineering using printing. *Biomaterials* **2009**, *30*, 5910–5917. [CrossRef] [PubMed]
13. Kaushik, S.N.; Kim, B.; Walma, A.M.; Choi, S.C.; Wu, H.; Mao, J.J.; Jun, H.W.; Cheon, K. Biomimetic microenvironments for regenerative endodontics. *Biomater. Res.* **2016**, *20*, 1–12. [CrossRef] [PubMed]
14. Holzl, K.; Lin, S.; Tytgat, L.; Van Vlierberghe, S.; Gu, L.; Ovsianikov, A. Bioink properties before, during and after 3D printing. *Biofabrication* **2016**, *8*, 032002. [CrossRef] [PubMed]
15. Hospodiuk, M.; Dey, M.; Sosnoski, D.; Ozbolat, I.T. The bioink: A comprehensive review on bioprintable materials. *Biotechnol. Adv.* **2017**, *35*, 217–239. [CrossRef] [PubMed]
16. Vepari, C.; Kaplan, D.L. Silk as a Biomaterial. *Prog. Polym. Sci.* **2007**, *32*, 991–1007. [CrossRef] [PubMed]
17. Yao, D.; Dong, S.; Lu, Q.; Hu, X.; Kaplan, D.L.; Zhang, B.; Zhu, H. Salt-leached silk scaffolds with tunable mechanical properties. *Biomacromolecules* **2012**, *13*, 3723–3729. [CrossRef] [PubMed]
18. Zhu, Z.H.; Ohgo, K.; Asakura, T. Preparation and characterization of regenerated Bombyx mori silk fibroin fiber with high strength. *Express Polym. Lett.* **2008**, *2*, 885–889. [CrossRef]
19. Liang, H.C.; Chang, W.H.; Liang, H.F.; Lee, M.H.; Sung, H.W. Crosslinking structures of gelatin hydrogels crosslinked with genipin or a water-soluble carbodiimide. *J. Appl. Polym. Sci.* **2004**, *91*, 4017–4026. [CrossRef]

20. Gruene, M.; Pflaum, M.; Hess, C.; Diamantouros, S.; Schlie, S.; Deiwick, A.; Koch, L.; Wilhelmi, M.; Jockenhoevel, S.; Haverich, A.; et al. Laser printing of three-dimensional multicellular arrays for studies of cell-cell and cell-environment interactions. *Tissue Eng. Part C Methods* **2011**, *17*, 973–982. [CrossRef]
21. Zhang, Y.; Yu, Y.; Chen, H.; Ozbolat, I.T. Characterization of printable cellular micro-fluidic channels for tissue engineering. *Biofabrication* **2013**, *5*, 025004. [CrossRef]
22. Zhang, Y.; Yu, Y.; Ozbolat, I.T. Direct Printing of Vessel-Like Tubular Microfluidic Channels. *J. Nanotechnol. Eng. Med.* **2013**, *4*, 2. [CrossRef]
23. Burdick, J.A.; Prestwich, G.D. Hyaluronic acid hydrogels for biomedical applications. *Adv. Mater.* **2011**, *23*, H41–H56. [CrossRef] [PubMed]
24. Moon, B.M.; Choi, M.J.; Sultan, M.T.; Yang, J.W.; Ju, H.W.; Lee, J.M.; Park, H.J.; Park, Y.R.; Kim, S.H.; Kim, D.W.; et al. Novel fabrication method of the peritoneal dialysis filter using silk fibroin with urease fixation system. *J. Biomed. Mater. Res. B Appl. Biomater.* **2017**, *105*, 2136–2144. [CrossRef]
25. Vasconcelos, A.; Gomes, A.C.; Cavaco-Paulo, A. Novel silk fibroin/elastin wound dressings. *Acta Biomater.* **2012**, *8*, 3049–3060. [CrossRef]
26. Gao, F.; Xu, Z.; Liang, Q.; Liu, B.; Li, H.; Wu, Y.; Zhang, Y.; Lin, Z.; Wu, M.; Ruan, C.; et al. Direct 3D Printing of High Strength Biohybrid Gradient Hydrogel Scaffolds for Efficient Repair of Osteochondral Defect. *Adv. Funct. Mater.* **2018**, *28*, 1706644. [CrossRef]
27. Garcia-Fuentes, M.; Meinel, A.J.; Hilbe, M.; Meinel, L.; Merkle, H.P. Silk fibroin/hyaluronan scaffolds for human mesenchymal stem cell culture in tissue engineering. *Biomaterials* **2009**, *30*, 5068–5076. [CrossRef]
28. Porter, D.; Vollrath, F. Silk as a Biomimetic Ideal for Structural Polymers. *Adv. Mater.* **2009**, *21*, 487–492. [CrossRef]
29. Brenckle, M.A.; Tao, H.; Kim, S.; Paquette, M.; Kaplan, D.L.; Omenetto, F.G. Protein-protein nanoimprinting of silk fibroin films. *Adv. Mater.* **2013**, *25*, 2409–2414. [CrossRef]
30. Guo, N.; Leu, M.C. Additive manufacturing: Technology, applications and research needs. *Front. Mech. Eng.* **2013**, *8*, 215–243. [CrossRef]
31. Kruth, J.P.; Leu, M.C.; Nakagawa, T. Progress in Additive Manufacturing and Rapid Prototyping. *Ann. CIRP* **1998**, *47*, 525–540. [CrossRef]
32. Webb, P.A. A review of rapid prototyping (RP) techniques in the medical and biomedical sector. *J. Med. Eng. Technol.* **2000**, *24*, 149–153. [CrossRef] [PubMed]
33. Jose, R.R.; Brown, J.E.; Polido, K.E.; Omenetto, F.G.; Kaplan, D.L. Polyol-Silk Bioink Formulations as Two-Part Room-Temperature Curable Materials for 3D Printing. *ACS Biomater. Sci. Eng.* **2015**, *1*, 780–788. [CrossRef]
34. Sommer, M.R.; Schaffner, M.; Carnelli, D.; Studart, A.R. 3D Printing of Hierarchical Silk Fibroin Structures. *ACS Appl. Mater. Interfaces* **2016**, *8*, 34677–34685. [CrossRef] [PubMed]
35. Rodriguez, M.J.; Brown, J.; Giordano, J.; Lin, S.J.; Omenetto, F.G.; Kaplan, D.L. Silk based bioinks for soft tissue reconstruction using 3-dimensional (3D) printing with in vitro and in vivo assessments. *Biomaterials* **2017**, *117*, 105–115. [CrossRef] [PubMed]
36. Kim, S.H.; Yeon, Y.K.; Lee, J.M.; Chao, J.R.; Lee, Y.J.; Seo, Y.B.; Sultan, M.T.; Lee, O.J.; Lee, J.S.; Yoon, S.I.; et al. Precisely printable and biocompatible silk fibroin bioink for digital light processing 3D printing. *Nat. Commun.* **2018**, *9*, 1620.
37. Rockwood, D.N.; Preda, R.C.; Yucel, T.; Wang, X.; Lovett, M.L.; Kaplan, D.L. Materials fabrication from *Bombyx mori* silk fibroin. *Nat. Protoc.* **2011**, *6*, 1612–1631. [CrossRef] [PubMed]
38. Ling, S.; Zhang, Q.; Kaplan, D.L.; Omenetto, F.; Buehler, M.J.; Qin, Z. Printing of stretchable silk membranes for strain measurements. *Lab Chip* **2016**, *16*, 2459–2466. [CrossRef] [PubMed]
39. Zhang, F.; You, X.; Dou, H.; Liu, Z.; Zuo, B.; Zhang, X. Facile fabrication of robust silk nanofibril films via direct dissolution of silk in $CaCl_2$-formic acid solution. *ACS Appl. Mater. Interfaces* **2015**, *7*, 3352–3361. [CrossRef] [PubMed]
40. Chameettachal, S.; Midha, S.; Ghosh, S. Regulation of Chondrogenesis and Hypertrophy in Silk Fibroin-Gelatin-Based 3D Bioprinted Constructs. *ACS Biomater. Sci. Eng.* **2016**, *2*, 1450–1463. [CrossRef]
41. Yeo, M.; Lee, J.S.; Chun, W.; Kim, G.H. An Innovative Collagen-Based Cell-Printing Method for Obtaining Human Adipose Stem Cell-Laden Structures Consisting of Core-Sheath Structures for Tissue Engineering. *Biomacromolecules* **2016**, *17*, 1365–1375. [CrossRef]

42. Diamantides, N.; Wang, L.; Pruiksma, T.; Siemiatkoski, J.; Dugopolski, C.; Shortkroff, S.; Kennedy, S.; Bonassar, L.J. Correlating rheological properties and printability of collagen bioinks: The effects of riboflavin photocrosslinking and pH. *Biofabrication* **2017**, *9*, 034102. [CrossRef] [PubMed]
43. Midha, S.; Murab, S.; Ghosh, S. Osteogenic signaling on silk-based matrices. *Biomaterials* **2016**, *97*, 133–153. [CrossRef] [PubMed]
44. Li, M.; Ogiso, M.; Minoura, N. Enzymatic degradation behavior of porous silk fibroin sheets. *Biomaterials* **2003**, *24*, 357–365. [CrossRef]
45. Mobini, S.; Hoyer, B.; Solati-Hashjin, M.; Lode, A.; Nosoudi, N.; Samadikuchaksaraei, A.; Gelinsky, M. Fabrication and characterization of regenerated silk scaffolds reinforced with natural silk fibers for bone tissue engineering. *J. Biomed. Mater. Res. A* **2013**, *101*, 2392–2404. [CrossRef] [PubMed]
46. Li, X.; Yan, S.; Qu, J.; Li, M.; Ye, D.; You, R.; Zhang, Q.; Wang, D. Soft freezing-induced self-assembly of silk fibroin for tunable gelation. *Int. J. Biol. Macromol.* **2018**, *117*, 691–695. [CrossRef] [PubMed]
47. Rossle, M.; Panine, P.; Urban, V.S.; Riekel, C. Structural evolution of regenerated silk fibroin under shear: Combined wide- and small-angle X-ray scattering experiments using synchrotron radiation. *Biopolymers* **2004**, *74*, 316–327. [CrossRef] [PubMed]
48. Im, D.S.; Kim, M.H.; Yoon, Y.I.; Park, W.H. Gelation Behaviors and Mechanism of Silk Fibroin According to the Addition of Nitrate Salts. *Int. J. Mol. Sci.* **2016**, *17*, 10. [CrossRef] [PubMed]
49. Terry, A.E.; Knight, D.P.; Porter, D.; Vollrath, F. pH Induced Changes in the Rheology of Silk Fibroin Solution from the Middle Division of *Bombyx mori* Silkworm. *Biomacromolecules* **2004**, *5*, 768–772. [CrossRef]
50. Das, S.; Pati, F.; Choi, Y.J.; Rijal, G.; Shim, J.H.; Kim, S.W.; Ray, A.R.; Cho, D.W.; Ghosh, S. Bioprintable, cell-laden silk fibroin-gelatin hydrogel supporting multilineage differentiation of stem cells for fabrication of three-dimensional tissue constructs. *Acta Biomater.* **2015**, *11*, 233–246. [CrossRef]
51. Sun, L.; Parker, S.T.; Syoji, D.; Wang, X.; Lewis, J.A.; Kaplan, D.L. Direct-write assembly of 3D silk/hydroxyapatite scaffolds for bone co-cultures. *Adv. Healthc. Mater.* **2012**, *1*, 729–735. [CrossRef]
52. Lee, H.; Yang, G.H.; Kim, M.; Lee, J.; Huh, J.; Kim, G. Fabrication of micro/nanoporous collagen/dECM/silk-fibroin biocomposite scaffolds using a low temperature 3D printing process for bone tissue regeneration. *Mater. Sci. Eng. C Mater. Biol. Appl.* **2018**, *84*, 140–147. [CrossRef] [PubMed]
53. Wu, Y.; Lin, Z.Y.; Wenger, A.C.; Tam, K.C.; Tang, X. 3D printing of liver-mimetic construct with alginate/cellulose nanocrystal hybrid bioink. *Printing* **2018**, *9*, 1–6.
54. Wei, J.; Wang, J.; Su, S.; Wang, S.; Qiu, J.; Zhang, Z.; Christopher, G.; Ning, F.; Cong, W. 3D printing of an extremely tough hydrogel. *RSC Adv.* **2015**, *5*, 81324–81329. [CrossRef]
55. Lee, J.; Yeo, M.; Kim, W.; Koo, Y.; Kim, G.H. Development of a tannic acid cross-linking process for obtaining 3D porous cell-laden collagen structure. *Int. J. Biol. Macromol.* **2018**, *110*, 497–503. [CrossRef] [PubMed]
56. Cui, X.; Boland, T. Human microvasculature fabrication using thermal inkjet printing technology. *Biomaterials* **2009**, *30*, 6221–6227. [CrossRef]
57. Markstedt, K.; Mantas, A.; Tournier, I.; Martinez Avila, H.; Hagg, D.; Gatenholm, P. 3D Printing Human Chondrocytes with Nanocellulose-Alginate Bioink for Cartilage Tissue Engineering Applications. *Biomacromolecules* **2015**, *16*, 1489–1496. [CrossRef]
58. Markstedt, K.; Escalante, A.; Toriz, G.; Gatenholm, P. Biomimetic Inks Based on Cellulose Nanofibrils and Cross-Linkable Xylans for 3D Printing. *ACS Appl. Mater. Interfaces* **2017**, *9*, 40878–40886. [CrossRef]
59. Ouyang, L.; Highley, C.B.; Rodell, C.B.; Sun, W.; Burdick, J.A. 3D Printing of Shear-Thinning Hyaluronic Acid Hydrogels with Secondary Cross-Linking. *ACS Sustain. Chem. Eng.* **2016**, *2*, 1743–1751. [CrossRef]
60. Ting, H.; Chunquan, F.; Min, Z.; Yufang, Z.; Weizhong, Z.; Lei, L. 3D-printed scaffolds of biomineralized hydroxyapatite nanocomposite on silk fibroin for improving bone regeneration. *Appl. Surf. Sci.* **2018**. [CrossRef]
61. Wang, Q.; Xia, Q.; Wu, Y.; Zhang, X.; Wen, F.; Chen, X.; Zhang, S.; Heng, B.C.; He, Y.; Ouyang, H.W. 3D-Printed Atsttrin-Incorporated Alginate/Hydroxyapatite Scaffold Promotes Bone Defect Regeneration with TNF/TNFR Signaling Involvement. *Adv. Healthc. Mater.* **2015**, *4*, 1701–1708. [CrossRef]
62. Kesti, M.; Muller, M.; Becher, J.; Schnabelrauch, M.; D'Este, M.; Eglin, D.; Zenobi-Wong, M. A versatile bioink for three-dimensional printing of cellular scaffolds based on thermally and photo-triggered tandem gelation. *Acta Biomater.* **2015**, *11*, 162–172. [CrossRef] [PubMed]
63. Gao, T.; Gillispie, G.J.; Copus, J.S.; Pr, A.K.; Seol, Y.J.; Atala, A.; Yoo, J.J.; Lee, S.J. Optimization of gelatin-alginate composite bioink printability using rheological parameters: A systematic approach. *Biofabrication* **2018**, *10*, 034106. [CrossRef] [PubMed]

64. Discher, D.E.; Janmey, P.; Wang, Y.L. Tissue Cells Feel and Respond to the Stiffness of Their Substrate. *Science* **2005**, *310*, 1139–1143. [CrossRef] [PubMed]
65. Knowles, T.P.; Buehler, M.J. Nanomechanics of functional and pathological amyloid materials. *Nat. Nanotechnol.* **2011**, *6*, 469–479. [CrossRef] [PubMed]
66. Tao, H.; Marelli, B.; Yang, M.; An, B.; Onses, M.S.; Rogers, J.A.; Kaplan, D.L.; Omenetto, F.G. Inkjet Printing of Regenerated Silk Fibroin: From Printable Forms to Printable Functions. *Adv. Mater.* **2015**, *27*, 4273–4279. [CrossRef] [PubMed]
67. Hodgkinson, T.; Chen, Y.; Bayat, A.; Yuan, X.F. Rheology and electrospinning of regenerated *Bombyx mori* silk fibroin aqueous solutions. *Biomacromolecules* **2014**, *15*, 1288–1298. [CrossRef] [PubMed]
68. Pan, H.; Zhang, Y.; Hang, Y.; Shao, H.; Hu, X.; Xu, Y.; Feng, C. Significantly reinforced composite fibers electrospun from silk fibroin/carbon nanotube aqueous solutions. *Biomacromolecules* **2012**, *13*, 2859–2867. [CrossRef]
69. Laity, P.R.; Holland, C. Native Silk Feedstock as a Model Biopolymer: A Rheological Perspective. *Biomacromolecules* **2016**, *17*, 2662–2671. [CrossRef]
70. Berman, B. 3-D printing: The new industrial revolution. *Bus. Horiz.* **2012**, *55*, 155–162. [CrossRef]
71. Murphy, A.R.; Kaplan, D.L. Biomedical applications of chemically-modified silk fibroin. *J. Mater. Chem.* **2009**, *19*, 6443–6450. [CrossRef]
72. Osman Rathore, D.Y.S. Nanostructure Formation through β-Sheet Self-Assembly in Silk-Based materials. *Macromolecules* **2001**, *34*, 1477–1486. [CrossRef]
73. Murphy, A.R.; John, P.S.; Kaplan, D.L. Corrigendum to 'Modification of silk fibroin using diazonium coupling chemistry and the effects on hMSC proliferation and differentiation' [Biomaterials 29 (2008) 2829–2838]. *Biomaterials* **2008**, *29*, 4260. [CrossRef]
74. Tamada, Y. Sulfation of silk fibroin by chlorosulfonic acid and the anticoagulant activity. *Biomaterials* **2004**, *25*, 377–383. [CrossRef]
75. Freddi, G.; Anghileri, A.; Sampaio, S.; Buchert, J.; Monti, P.; Taddei, P. Tyrosinase-catalyzed modification of *Bombyx mori* silk fibroin: Grafting of chitosan under heterogeneous reaction conditions. *J. Biotechnol.* **2006**, *125*, 281–294. [CrossRef] [PubMed]
76. Costa, J.B.; Silva-Correia, J.; Oliveira, J.M.; Reis, R.L. Fast Setting Silk Fibroin Bioink for Printing of Patient-Specific Memory-Shape Implants. *Adv. Healthc. Mater.* **2017**, *6*, 1701021. [CrossRef] [PubMed]
77. Chimene, D.; Lennox, K.K.; Kaunas, R.R.; Gaharwar, A.K. Advanced Bioinks for 3D Printing: A Materials Science Perspective. *Ann. Biomed. Eng.* **2016**, *44*, 2090–2102. [CrossRef]
78. Compaan, A.M.; Christensen, K.; Huang, Y. Inkjet Printing of 3D Silk Fibroin Cellular Constructs Using Sacrificial Alginate. *ACS Biomater. Sci. Eng.* **2016**, *3*, 1519–1526. [CrossRef]
79. Xiong, S.; Zhang, X.; Lu, P.; Wu, Y.; Wang, Q.; Sun, H.; Heng, B.C.; Bunpetch, V.; Zhang, S.; Ouyang, H. A Gelatin-sulfonated Silk Composite Scaffold based on 3D Printing Technology Enhances Skin Regeneration by Stimulating Epidermal Growth and Dermal Neovascularization. *Sci. Rep.* **2017**, *7*, 4288. [CrossRef]
80. Meinel, L.; Kaplan, D.L. Silk constructs for delivery of musculoskeletal therapeutics. *Adv. Drug Deliv. Rev.* **2012**, *64*, 1111–1122. [CrossRef]
81. Bandyopadhyay, A.; Bose, S.; Das, S. 3D printing of biomaterials. *MRS Bull.* **2015**, *40*, 108–115. [CrossRef]
82. Shim, J.H.; Kim, J.Y.; Park, M.; Park, J.; Cho, D.W. Development of a hybrid scaffold with synthetic biomaterials and hydrogel using solid freeform fabrication technology. *Biofabrication* **2011**, *3*, 034102. [CrossRef] [PubMed]
83. Hong, S.; Sycks, D.; Chan, H.F.; Lin, S.; Lopez, G.P.; Guilak, F.; Leong, K.W.; Zhao, X. 3D Printing of Highly Stretchable and Tough Hydrogels into Complex, Cellularized Structures. *Adv. Mater.* **2015**, *27*, 4035–4040. [CrossRef] [PubMed]
84. Kapoor, S.; Kundu, S.C. Silk protein-based hydrogels: Promising advanced materials for biomedical applications. *Acta Biomater.* **2016**, *31*, 17–32. [CrossRef] [PubMed]
85. Chao, P.H.; Yodmuang, S.; Wang, X.; Sun, L.; Kaplan, D.L.; Vunjak-Novakovic, G. Silk hydrogel for cartilage tissue engineering. *J. Biomed. Mater. Res. B Appl. Biomater.* **2010**, *95*, 84–90. [CrossRef] [PubMed]
86. Han, L.; Sun, H.; Tang, P.; Li, P.; Xie, C.; Wang, M.; Wang, K.; Weng, J.; Tan, H.; Ren, F.; et al. Mussel-inspired graphene oxide nanosheet-enwrapped Ti scaffolds with drug-encapsulated gelatin microspheres for bone regeneration. *Biomater. Sci.* **2018**, *6*, 538–549. [CrossRef] [PubMed]

87. Shi, W.; Sun, M.; Hu, X.; Ren, B.; Cheng, J.; Li, C.; Duan, X.; Fu, X.; Zhang, J.; Chen, H.; et al. Structurally and Functionally Optimized Silk-Fibroin-Gelatin Scaffold Using 3D Printing to Repair Cartilage Injury In Vitro and In Vivo. *Adv. Mater.* **2017**, *29*, 1701089. [CrossRef] [PubMed]
88. Hong, N.; Yang, G.H.; Lee, J.; Kim, G. 3D printing and its in vivo applications. *J. Biomed. Mater. Res. B Appl. Biomater.* **2018**, *106*, 444–459. [CrossRef]
89. Highley, C.B.; Rodell, C.B.; Burdick, J.A. Direct 3D Printing of Shear-Thinning Hydrogels into Self-Healing Hydrogels. *Adv. Mater.* **2015**, *27*, 5075–5079. [CrossRef]
90. Li, Z.; Jia, S.; Xiong, Z.; Long, Q.; Yan, S.; Hao, F.; Liu, J.; Yuan, Z. 3D-printed scaffolds with calcified layer for osteochondral tissue engineering. *J. Biosci. Bioeng.* **2018**, *126*, 389–396. [CrossRef]
91. Cui, X.; Boland, T.; D'Lima, D.; Lotz, M. Thermal Inkjet Printing in Tissue Engineering and Regenerative Medicine. *Recent Pat. Drug Deliv. Formul.* **2012**, *6*, 149–155. [CrossRef]
92. Malda, J.; Visser, J.; Melchels, F.P.; Jungst, T.; Hennink, W.E.; Dhert, W.J.; Groll, J.; Hutmacher, D.W. 25th anniversary article: Engineering hydrogels for biofabrication. *Adv. Mater.* **2013**, *25*, 5011–5028. [CrossRef]
93. Rujiravanit, R.; Kruaykitanon, S.; Jamieson, A.M.; Tokura, S. Preparation of Crosslinked Chitosan/Silk Fibroin Blend Films for Drug Delivery System. *Macromol. Biosci.* **2003**, *3*, 604–611. [CrossRef]
94. Chrisey, D.B.; Pique, A.; Fitz-Gerald, J.; Auyeung, R.C.Y.; McGill, R.A.; Wu, H.D.; Duignan, M. New approach to laser direct writing active and passive. *Appl. Surf. Sci.* **1999**, *154*, 593–600.
95. Duan, B.; Hockaday, L.A.; Kang, K.H.; Butcher, J.T. 3D printing of heterogeneous aortic valve conduits with alginate/gelatin hydrogels. *J. Biomed. Mater. Res. A* **2013**, *101*, 1255–1264. [CrossRef] [PubMed]
96. Murphy, S.V.; Atala, A. 3D printing of tissues and organs. *Nat. Biotechnol.* **2014**, *32*, 773–785. [CrossRef] [PubMed]
97. Zheng, Z.; Wu, J.; Liu, M.; Wang, H.; Li, C.; Rodriguez, M.J.; Li, G.; Wang, X.; Kaplan, D.L. 3D Printing of Self-Standing Silk-Based Bioink. *Adv. Healthc. Mater.* **2018**, *7*, e1701026. [CrossRef] [PubMed]
98. Zhang, J.; Allardyce, B.J.; Rajkhowa, R.; Zhao, Y.; Dilley, R.J.; Redmond, S.L.; Wang, X.; Liu, X. 3D Printing of Silk Particle-Reinforced Chitosan Hydrogel Structures and Their Properties. *ACS Biomater. Sci. Eng.* **2018**, *4*, 3036–3046. [CrossRef]
99. Rider, P.; Zhang, Y.; Tse, C.; Zhang, Y.; Jayawardane, D.; Stringer, J.; Callaghan, J.; Brook, I.M.; Miller, C.A.; Zhao, X.; et al. Biocompatible silk fibroin scaffold prepared by reactive inkjet printing. *J. Mater. Sci.* **2016**, *51*, 8625–8630. [CrossRef]
100. Catros, S.; Guillemot, F.; Nandakumar, A.; Ziane, S.; Moroni, L.; Habibovic, P.; van Blitterswijk, C.; Rousseau, B.; Chassande, O.; Amedee, J.; et al. Layer-by-layer tissue microfabrication supports cell proliferation in vitro and in vivo. *Tissue Eng. Part C Methods* **2012**, *18*, 62–70. [CrossRef]
101. Pati, F.; Ha, D.H.; Jang, J.; Han, H.H.; Rhie, J.W.; Cho, D.W. Biomimetic 3D tissue printing for soft tissue regeneration. *Biomaterials* **2015**, *62*, 164–175. [CrossRef] [PubMed]
102. Bouwstra, J.A.; Honeywell-Nguyen, P.L. Skin structure and mode of action of vesicles. *Adv. Drug Deliv. Rev.* **2002**, *54*, S41–S45. [CrossRef]
103. Khalili, S.; Khorasani, S.N.; Razavi, S.M.; Hashemibeni, B.; Tamayol, A. Nanofibrous Scaffolds with Biomimetic Composition for Skin Regeneration. *Appl. Biochem. Biotechnol.* **2018**. [CrossRef] [PubMed]
104. Yeo, I.S.; Oh, J.E.; Jeong, L.; Lee, T.S.; Lee, S.J.; Park, W.H.; Min, B.M. Collagen-Based Biomimetic Nanofibrous Scaffolds—Preparation andcharacterization of collagen&silk fibroin bicomponent nanofibrous structures. *Biomacromolecules* **2008**, *9*, 1106–1116. [PubMed]
105. Loeser, R.F. Age-related changes in the musculoskeletal system and the development of osteoarthritis. *Clin. Geriatr. Med.* **2010**, *26*, 371–386. [CrossRef] [PubMed]
106. Do, A.V.; Khorsand, B.; Geary, S.M.; Salem, A.K. 3D Printing of Scaffolds for Tissue Regeneration Applications. *Adv. Healthc. Mater.* **2015**, *4*, 1742–1762. [CrossRef]
107. Dababneh, A.B.; Ozbolat, I.T. Printing Technology: A Current State-of-the-Art Review. *J. Manuf. Sci. Eng.* **2014**, *136*, 061016. [CrossRef]
108. Chawla, S.; Midha, S.; Sharma, A.; Ghosh, S. Silk-Based Bioinks for 3D Printing. *Adv. Healthc. Mater.* **2018**, *7*, e1701204. [CrossRef]
109. Melke, J.; Midha, S.; Ghosh, S.; Ito, K.; Hofmann, S. Silk fibroin as biomaterial for bone tissue engineering. *Acta Biomater.* **2016**, *31*, 1–16. [CrossRef]
110. Stevens, M.M. Biomaterials for bone tissue engineering. *Mater. Today* **2008**, *11*, 18–25. [CrossRef]

111. Alehosseini, M.; Golafshan, N.; Kharaziha, M. Design and characterization of poly-ε-caprolactone electrospun fibers incorporated with α-TCP nanopowder as a potential guided bone regeneration membrane. *Mater. Today Proc.* **2018**, *5*, 15783–15789. [CrossRef]
112. Shkarina, S.; Shkarin, R.; Weinhardt, V.; Melnik, E.; Vacun, G.; Kluger, P.; Loza, K.; Epple, M.; Ivlev, S.I.; Baumbach, T.; et al. 3D biodegradable scaffolds of polycaprolactone with silicate-containing hydroxyapatite microparticles for bone tissue engineering: High-resolution tomography and in vitro study. *Sci. Rep.* **2018**, *8*, 8907. [CrossRef]
113. Gao, G.; Lee, J.H.; Jang, J.; Lee, D.H.; Kong, J.-S.; Kim, B.S.; Choi, Y.-J.; Jang, W.B.; Hong, Y.J.; Kwon, S.-M.; et al. Tissue Engineered Bio-Blood-Vessels Constructed Using a Tissue-Specific Bioink and 3D Coaxial Cell Printing Technique: A Novel Therapy for Ischemic Disease. *Adv. Funct. Mater.* **2017**, *27*, 1700798. [CrossRef]
114. Adler-Abramovich, L.; Arnon, Z.A.; Sui, X.; Azuri, I.; Cohen, H.; Hod, O.; Kronik, L.; Shimon, L.J.W.; Wagner, H.D.; Gazit, E. Bioinspired Flexible and Tough Layered Peptide Crystals. *Adv. Mater.* **2018**, *30*, 1704551. [CrossRef] [PubMed]
115. Jeyaraj, R.; Natasha, G.; Kirby, G.; Rajadas, J.; Mosahebi, A.; Seifalian, A.M.; Tan, A. Vascularisation in regenerative therapeutics and surgery. *Mater. Sci. Eng. C Mater. Biol. Appl.* **2015**, *54*, 225–238. [CrossRef] [PubMed]
116. Shin, S.; Kwak, H.; Hyun, J. Melanin Nanoparticle-Incorporated Silk Fibroin Hydrogels for the Enhancement of Printing Resolution in 3D-Projection Stereolithography of Poly(ethylene glycol)-Tetraacrylate Bio-ink. *ACS Appl. Mater. Interfaces* **2018**, *10*, 23573–23582. [CrossRef] [PubMed]
117. Zhang, W.J.; Liu, W.; Cui, L.; Cao, Y. Tissue engineering of blood vessel. *J. Cell Mol. Med.* **2007**, *11*, 945–957. [CrossRef] [PubMed]
118. Unger, R.E.; Ghanaati, S.; Orth, C.; Sartoris, A.; Barbeck, M.; Halstenberg, S.; Motta, A.; Migliaresi, C.; Kirkpatrick, C.J. The rapid anastomosis between prevascularized networks on silk fibroin scaffolds generated in vitro with cocultures of human microvascular endothelial and osteoblast cells and the host vasculature. *Biomaterials* **2010**, *31*, 6959–6967. [CrossRef]

© 2019 by the authors. Licensee MDPI, Basel, Switzerland. This article is an open access article distributed under the terms and conditions of the Creative Commons Attribution (CC BY) license (http://creativecommons.org/licenses/by/4.0/).

Article

The Shape-Memory Effect of Hindered Phenol (AO-80)/Acrylic Rubber (ACM) Composites with Tunable Transition Temperature

Shi-kai Hu [1,2], Si Chen [1], Xiu-ying Zhao [1,*], Ming-ming Guo [1,2] and Li-qun Zhang [1]

[1] Key Laboratory of Beijing City on Preparation and Processing of Novel Polymer Materials, Beijing University of Chemical Technology, Beijing 100029, China; 2017400073@mail.buct.edu.cn (S.-k.H.); aircs123@163.com (S.C.); guomm57@swu.edu.cn (M.-m.G.); zhanglq@mail.buct.edu.cn (L.-q.Z.)
[2] SINOPEC Beijing Research Institute of Chemical Industry, Beijing 100013, China
* Correspondence: zhaoxy@mail.buct.edu.cn; Tel.: +86-10-6443-4860

Received: 31 October 2018; Accepted: 26 November 2018; Published: 4 December 2018

Abstract: To broaden the types and scope of use of shape-memory polymers (SMPs), we added the hindered phenol 3,9-bis[1,1-dimethyl-2-{b-(3-tert-butyl-4-hydroxy-5-methylphenyl)propionyloxy}ethyl] -2,4,8,10-tetraoxaspiro-[5,5]-undecane (AO-80), which comprises small organic molecules, to acrylic rubber (ACM) to form a series of AO-80/ACM rubber composites. The structural, thermal, mechanical property, and shape-memory properties of the AO-80/ACM rubber composites were investigated. We identified the formation of intra-molecular hydrogen bonding between –OH of AO-80 and the carbonyl groups and the ether groups of ACM molecules. The amount of AO-80 used can be adjusted to tailor the transition temperature. AO-80/ACM rubber composites showed excellent shape recovery and fixity. The approach for adjusting the transition temperature of AO-80/ACM rubber composites provides remarkable ideas for the design and preparation of new SMPs.

Keywords: acrylic rubber; shape-memory polymer; hindered phenol; hydrogen bonding

1. Introduction

Shape-memory materials (SMMs) can change from one pre-determined shape to another in response to a certain stimulus [1,2]. Research on shape-memory polymers (SMPs) can be fundamental and applied. SMPs possess many advantages over their well-investigated metallic counterparts, shape-memory alloys; these advantages include excellent processability, light weight, and notable flexibility in terms of material design [3–5]; SMP applications include medical devices, actuators, sensors, artificial muscles, switches, smart textiles, and self-deployable structures [4–7]. SMPs can return into an original shape upon the application of stimuli, such as temperature [8–10], humidity [11,12], light [13–16], electricity [8,17–20], pH [15,21–24], and irradiation. This memory phenomenon is because a polymer network has reversible and fixed phases. The reversible phases can be shaped under certain conditions. Reversible phases use ionic bond [1,25], vitrification [25,26], reversible crystallization [27], hydrogen bond [28,29], or supramolecular interactions [30,31] to maintain this metastable shape until an activation energy is used to facilitate a return to the original shape. The fixed phases allow deformation but hold the relative location of the chains. Fixed phases include physical and covalent cross-links, such as crystalline or glassy domains in polymers, or supramolecular interactions [32]. For thermally induced SMPs, when the deformation of SMP is above its switch transition temperature (T_{trans}) and then cooled below T_{trans}, most internal stress can be stored in cross-linking structure; by heating the SMP above its T_{trans}, the SMP recovers its original shape by releasing the internal stress [33,34]. When reheated above T_{trans} without stress, the cross-linking phase assumes its permanent shape. T_{trans} can either be the glass transition temperature (T_g) or melting temperature (T_m) of polymers. In general, the temperature province of T_{trans} of current SMMs reaches above room temperature. However, in specific conditions, such as deep-sea and polar region explorations, T_{trans} of SMMs should be lower than room temperature and can be adjusted and controlled by specific methods. A critical parameter for SMPs lies in its shape memory T_{trans}. For an amorphous SMP polymer, it is important to develop new methods to tailor its T_g, which corresponds to its shape memory T_{trans}. Zhao et al. created a nano- or molecule-scale-hindered phenol and polar rubber compound. Their research indicated that T_g of the developed material could be tailored by changing the kind and dosage of small organic molecule-hindered phenol [35,36]. This phenomenon was attributed to hydrogen bonding between hindered phenol 3,9-bis[1,1-dimethyl-2-{b-(3-tert-butyl-4-hydroxy-5-methylphenyl)propionyloxy}ethyl]-2,4,8,10-tetraoxaspiro-[5,5]-undecane (AO-80) and polar rubber. Such interactions will result in the molecular-level dispersion of AO-80 in CPE and rubber matrix and enhancement of intermolecular friction, which will further increase T_g. It is well known that typical epoxy-based materials which have been applied extensively in coatings, adhesives, and matrix material for structural composites are rigid with relatively low failure strains. There are many references regarding shape-memory epoxy composites that all have good shape memory with a high shape fixity (R_f) ratio and high shape recovery ratio (R_r), but these composites all have a short elongation at break [37–42]. In this study, AO-80 had been studied to prepare AO-80/acrylic rubber (ACM) nanocomposites with high failure strains compared to shape-memory epoxy composites. The structure of AO-80 is shown in Figure 1. AO-80/ACM rubber nanocomposites possibly possess remarkable filler/matrix interfacial properties because the AO-80 molecule features numerous polar functional groups (hydroxyl and carbonyl) that can form strong intermolecular interactions with ACM. An elastomer will exhibit shape-memory functionality when the material can be stabilized in the deformed state in a temperature range that is relevant for particular applications. Similar to normal polymers, SMPs also possess 3D molecular network-like architectures. ACM can exhibit 3D network structures after crosslinking. These cross-linked structures ensure that the polymer can maintain a stable shape at the macroscopic level by enabling the original and recovered shapes. This system also features a T_g below the room temperature, and temperature can be adjusted and controlled within a particular scope by incorporating small organic molecules to increase T_g [35,36], which will broaden the kind and scope of use of SMPs. In this study, we designed a series of AO-80/ACM rubber composites with high failure strains, the T_{trans} of which can be tailored by adding a dosage of small organic molecule-hindered phenol. No study or similar work has investigated the shape-memory effect of AO-80/ACM rubber composites, thereby broadening the list of SMPs with excellent shape-memory properties.

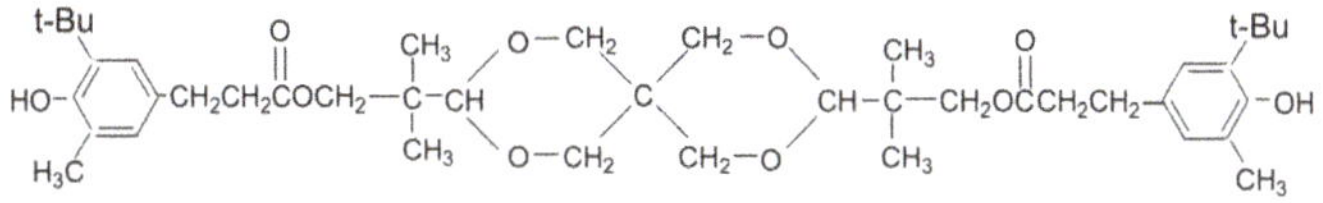

Figure 1. Chemical structure of hindered phenol 3,9-bis[1,1-dimethyl-2-{b-(3-tert-butyl-4-hydroxy-5-methylphenyl)propionyloxy}ethyl]-2,4,8,10-tetraoxaspiro-[5,5]-undecane (AO-80).

2. Materials and Methods

2.1. Materials

ACM (AR-801) was provided by Tohpe Corp (Sakai, Japan). AO-80 was obtained from Asahi Denka (Tokyo, Japan). Other ingredients and chemicals were obtained from China and were used as received.

2.2. Sample Preparations

AO-80/ACM rubber composites were obtained as follows: (1) After ACM was kneaded for 3 min, AO-80 (without previous treatment) was added into ACM. (2) After these mixtures were kneaded for 5 min, the AO-80/ACM mixtures were blended with compounding and crosslinking additives, including 5.0 phr of zinc oxide(CAS No:1314-13-2), 1.0 phr of stearic acid(CAS No: 57-11-4), 0.5 phr of potassium stearate(CAS No: 593-29-3), 4 phr of sodium stearate(CAS No: 822-16-2), and 0.5 phr of sulfur(CAS No: 7704-34-9). The mixtures were then kneaded for 10 min. The mixtures of AO-80/ACM were kept for at least 24 h. (3) Finally, the mixtures of AO-80/ACM were set at 180 °C and 15 MPa for 20 min and then naturally cooled down to prepare AO-80/ACM rubber composites.

2.3. Methods

The structure, shape-memory properties, and mechanical and thermal properties of AO-80/ACM rubber composites were systematically evaluated by differential scanning calorimetry (DSC), dynamic mechanical analysis (DMA), and Fourier-transform infrared (FT-IR) spectroscopy. The DSC curves were acquired from −60 °C to 150 °C at a rate of 10 °C/min with a STAR^e system calorimeter (Mettler–Toledo Co., Zurich, Switzerland). FT-IR spectra were acquired by using a Spectra-Tech ATR attachment to scan the samples.

The static mechanical properties of AO-80/ACM rubber composites were determined according to ASTM D638 by using a CMT4104 Electrical Tensile Tester (SANS Testing Machine Co., ShenZhen, China) at a rate of 500 mm/min at room temperature. The strip dimensions for testing were 20 mm in length, 6 mm in width, and 2 mm in thickness. Hardness was tested according to ASTM D2240-2015.

The shape-memory effect analysis of AO-80/ACM rubber composites was investigated on the DMA Q800 (TA Instruments, New Castle, DE, USA) using controlled-force mode with rectangular samples (6 mm in width and 2 mm in thickness). Prior to the investigation, the temperature was adjusted to an equilibration at T_{trans} + 20 °C for 10 min. In step 1 (deformation), the sample was stretched to a designed value (ε = 55%, ε = 100%, ε = 130%) by ramping the force from a preload value of 0.005 N at a rate of 0.5 N/min. In step 2 (cooling), the specimen was cooled to fix the deformed sample under constant force at the rate of 3 °C/min to T_{trans} − 20 °C. In step 3 (unloading and fixing), the force of the specimen was unloaded at a rate of 0.5 N/min to a preload value (0.005 N). Then, an equilibration at T_{trans} − 20 °C for 10 min to ensure shape fixing was performed. In the final step (recovery), the specimen was reheated to T_{trans} + 60 °C at the rate of 3 °C/min [37]. All experiments were carried out three times successively and the average results between second and third cycles are shown in the paper. From the curves, the shape recovery ratio (R_r) and the shape fixity ratio (R_f) for the shape-memory effect were computed as follows:

$$\text{Shape recovery}: R_r(N) = \frac{\varepsilon_m - \varepsilon_p(N)}{\varepsilon_m - \varepsilon_p(N-1)}, \times 100\% \tag{1}$$

$$\text{Shape fixity}: R_f(N) = \frac{\varepsilon_u(N)}{\varepsilon_m(N)} \times 100\% \tag{2}$$

where $\varepsilon_m, \varepsilon_u$ and ε_p are strains after the step of cooling, unloading, and recovery process, respectively. N refers to a consecutive number in a cyclic shape-memory measurement.

Dynamic mechanical properties were investigated on a DMA (Rheometric Scientific Co., Piscataway, NJ, USA). The strip dimensions for testing were 20 mm in length, 6 mm in width, and 2 mm in thickness. The curves of E'-T were acquired from −60 °C to 150 °C at a rate of 3 °C/min and with a frequency of 1 Hz at an amplitude of ε = 0.3%.

Shape recovery observations of the AO-80/ACM rubber composites were carried out in water. The composites were cut into rectangular strips with dimensions of 100.0 mm × 10.0 mm × 2.0 mm. The rectangular strips were fixed in a temporary shape at T_{high} and then cooled down to T_{low}. The rectangular strips in temporary shape were placed in a water bath at T_{high} while recording images of shape recovery using a video camera at a rate of 20 frames/s. Among the aforementioned procedure/conditions, T_{high} was equal to T_{trans} + 20 °C, and T_{low} was equal to T_{trans} − 20 °C.

3. Results

3.1. T_g of AO-80/ACM Rubber Composites

Figure 2 shows that the neat ACM featured a T_g of approximately −11 °C. Compared with the neat ACM, AO-80/ACM composites showed a T_g between those of neat ACM and quenched AO-80(40.9) [36]. T_g of AO-80/ACM rubber composites shifted from −11 °C to 10 °C when the dosage of AO-80 was added from zero phr to one hundred phr. The DSC curves of the composites showed neither T_g peak nor melting of AO-80 [36,43], which suggest that dispersion of AO-80 in ACM was at the molecular level by blending, and AO-80/ACM rubber composites were successfully prepared as expected. Strong intermolecular interactions were formed between AO-80 molecules and polar functional groups (ester and ether groups) of ACM. Hydrogen bonding between ACM and AO-80 are analyzed later. With both polar molecules, intermolecular interactions significantly hindered the slide of ACM chain and increased T_g of ACM composites.

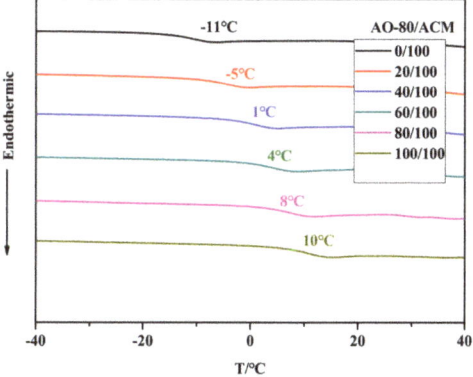

Figure 2. DSC curves of AO-80/acrylic rubber (ACM) rubber composites.

3.2. FT-IR of AO-80/ACM Rubber Composites

Interactions between different functional groups can be investigated through molecular dynamics simulation and FT-IR [44,45]. Figure 3 shows the FT-IR/ATR spectra of neat ACM and AO-80/ACM rubber composites. Figure 3a shows that the FT-IR/ATR spectra of all AO-80/ACM rubber composites indicate significantly wide peaks at 1135 cm^{-1} to 1195 cm^{-1}, which were assigned to C-O-C bending vibration and symmetric and antisymmetric stretching vibrations. The peak position gradually shifted to a higher wave number from 1158.5 cm^{-1} to 1163 cm^{-1} when the dosage of AO-80 was added from zero phr to one hundred phr, determining that -O- of C-O-C can bond with-OH of AO-80. Figure 3b shows the composition dependence of FT-IR spectra for the –C=O stretching regions of AO-80/ACM rubber composites. As AO-80 content increased, the –C=O peak position shifted to a higher wave number from 1730.0 cm^{-1} to 1732.0 cm^{-1} when the dosage of AO-80 was added from zero phr to one hundred phr. Studies reported that hydrogen-bonded vibration will present a frequency shift [35,36]. Figure 3c shows the –OH stretching regions of AO-80/ACM rubber composites. The position of –OH peak shifted to a lower wave number from 3555.1 cm^{-1} to 3498.7 cm^{-1} when the dosage of AO-80 was added from zero phr to one hundred phr. The hydrogen bonding between carbonyl and ether groups of segments of ACM and -OH groups of AO-80 was observed. The total frequency shift as a measure of the strength of hydrogen bonding is generally accepted [46–48]. Thus, these results indicate that as the dosage of AO-80 increased, the strength of the hydrogen bonding among functional groups between ACM and AO-80 improved. The result corroborates that the T_g of AO-80/ACM rubber composites increased with the dosage of AO-80, increasing because of hydrogen bonding. Figure 4 shows the possible hydrogen bonding of AO-80/ACM rubber composites.

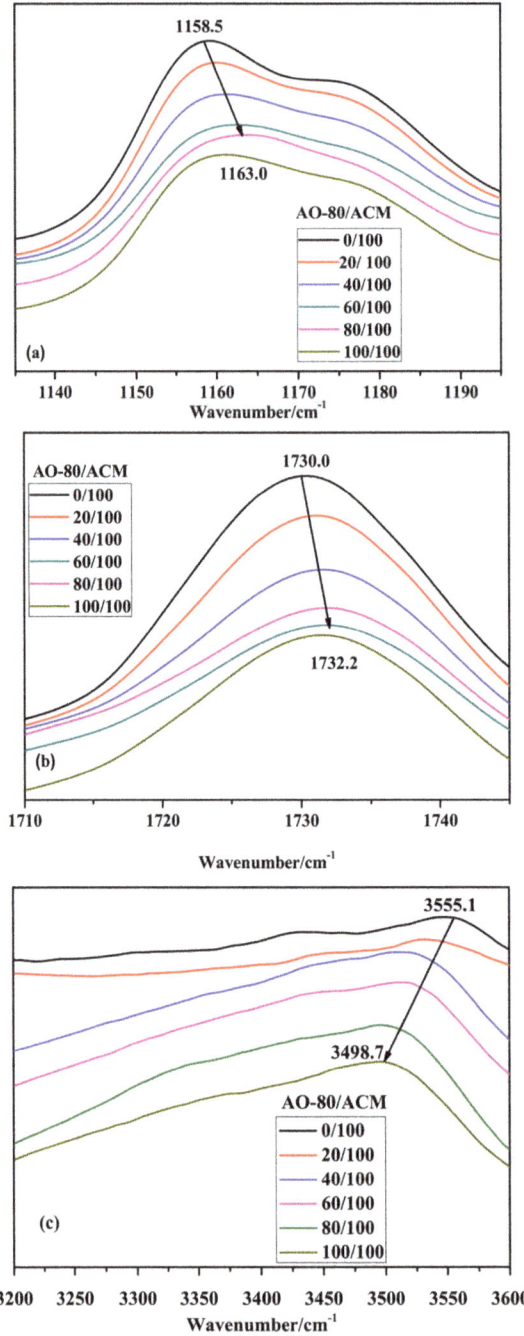

Figure 3. FT-IR spectra acquired at: (**a**) 1135 cm^{-1} to 1195 cm^{-1}; (**b**) 1710 cm^{-1} to 1745 cm^{-1}; and (**c**) 3200 cm^{-1} to 3600 cm^{-1} region for AO-80/ACM rubber composites.

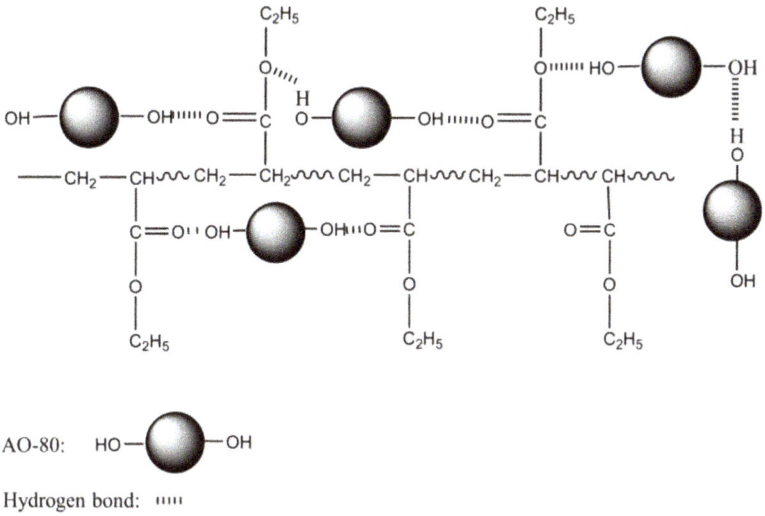

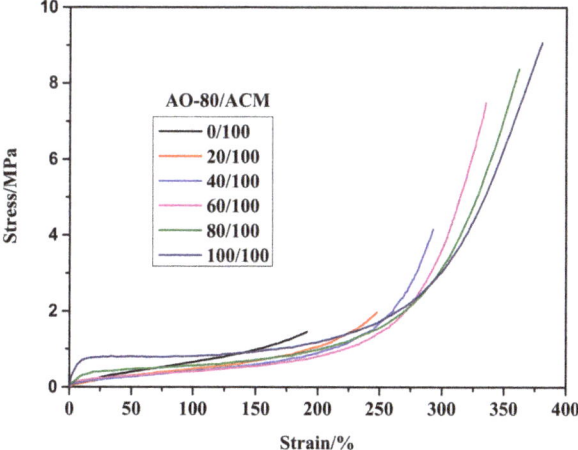

Figure 4. Possible hydrogen bond between AO-80 and ACM.

3.3. Static Mechanical Properties of AO-80/ACM Rubber Composites

The results of the tensile testing of neat ACM and AO-80/ACM rubber composites are shown in Figure 5 and the acquired data is summarized in Table 1. The elongation and tensile strength at break of the neat ACM were 210% and 1.47 MPa, respectively. All of the AO-80/ACM rubber composites with a content of AO-80 above forty phr had much longer elongation and higher tensile strength at break than ACM. This was because AO-80 had a reinforcement effect when AO-80 was added over 40 phr and the strength of hydrogen bonding among functional groups between ACM and AO-80 was improved when the AO-80 content was added increasingly.

Figure 5. Stress-strain curves of ACM and AO-80/ACM rubber composites.

Table 1. Mechanical properties of AO-80/ACM rubber composites.

Properties	Loadings of AO-80/phr					
	0	20	40	60	80	100
Hardness (Shore A)	41 ± 0	48 ± 0	68 ± 0	78 ± 0	93 ± 0	95 ± 0
Tensile strength (MPa)	1.5 ± 0.2	1.9 ± 0.1	4.0 ± 0.2	7.7 ± 0.1	8.2 ± 0.1	9.2 ± 0.2
Elongation at break (%)	210 ± 9	248 ± 11	295 ± 12	336 ± 8	369 ± 8	377 ± 5

3.4. Shape-Memory Effect of AO-80/ACM Rubber Composite

Figure 6 depicts the 3D ε-T-σ curves of various compositions for AO-80/ACM rubber composites. The results showed that the samples were generally further deformed because of loading during the cooling/fixing step after deformation, and the T_g of AO-80/ACM rubber composites increased with an increasing dosage of AO-80; in other words, the T_{trans} of AO-80/ACM rubber composites also increased with increasing AO-80. All samples exhibited excellent shape recovery, as shown in Figure 6. All the samples presented a high shape fixing ratio and recovery ratio when they were stretched to a given strain (100%). R_r and R_f were both above 99%. Figure 7 plots the 3D ε-T-σ curves of five cycles for AO-80/ACM (40/100) rubber composite. The 3D ε-T-σ curves of AO-80/ACM (40/100) rubber composites were similar with different cycles. Different cycles all showed high shape fixing and recovery rates. The results showed the repeatability of AO-80/ACM rubber composites as shape-memory materials were excellent. The excellent repeatability of AO-80/ACM rubber composites was due to good elasticity of samples. Figure 8 plots the 3D ε-T-σ curves of different strains (deformation) for AO-80/ACM (60/100) rubber composite. All the diagrams show high shape fixing and recovery ratio when the given strains were 55%, 100%, and 130%. R_r reached above 99%, and R_f was above 99%. The results show that the range of deformation for the AO-80/ACM rubber composites as shape-memory materials is broad, which is due to high elongation at break of AO-80/ACM rubber composites. Figure 9 displays the R_r-T curves of AO-80/ACM rubber composites with various compositions. A significant portion of prestrain was recovered in all samples within the temperature range of T_{10}–T_{90}. With increasing AO-80, the recovery temperature, T_{10} (R_r = 10%), T_{50} (R_r = 50%), T_{90} (R_r = 90%) increased, which was due to intermolecular interactions significantly hindering the slide of ACM chain and increasing the T_g (T_{trans}) of AO-80/ACM rubber composites. Figures 6–9 show that AO-80/ACM rubber composites exhibit excellent shape-memory behavior.

The possible molecular mechanism of AO-80/ACM rubber composites is that AO-80/ACM rubber composites consist of molecular switches that are temperature-sensitive netpoints. The permanent shape in AO-80/ACM rubber composites was determined by netpoints that are cross-linked by the cross-linking agent. The temporary shape was fixed by the vitrification of AO-80/ACM rubber composites. Samples can be deformed to a temporary shape above T_{trans} + 20 °C, and the shape can be fixed at T_{trans} − 20 °C under stress. When heated above T_{trans} + 60 °C without stress, the specimen recovered its original shape because of the netpoints.

Materials **2018**, *11*, 2461

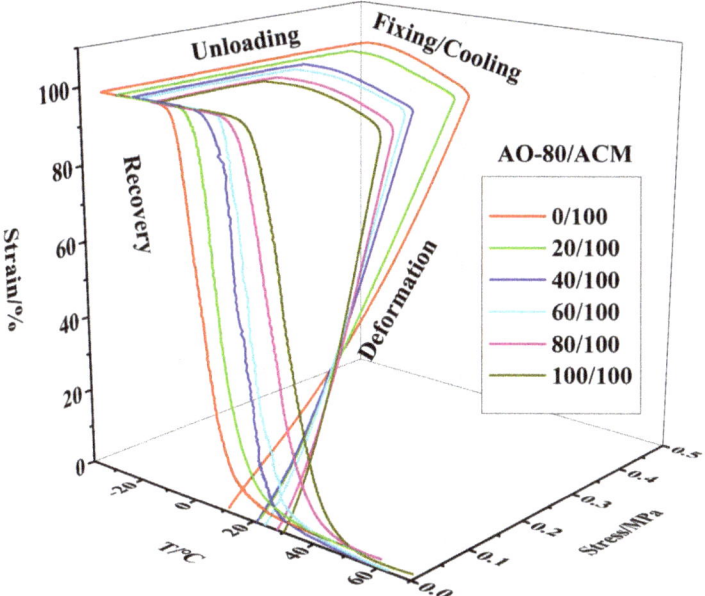

Figure 6. 3D ε-T-σ curve of various compositions for AO-80/ACM rubber composites.

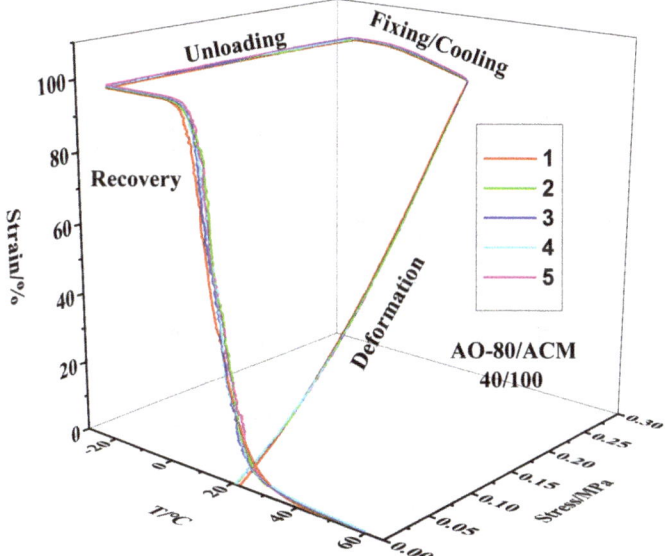

Figure 7. 3D ε-T-σ curve of five cycles for AO-80/ACM (40/100) rubber composite.

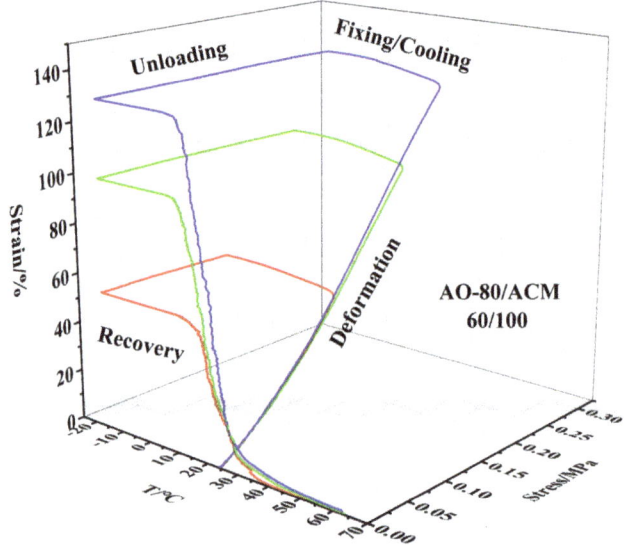

Figure 8. 3D ε-T-σ curves of different strains (deformation) for AO-80/ACM (60/100) rubber composite.

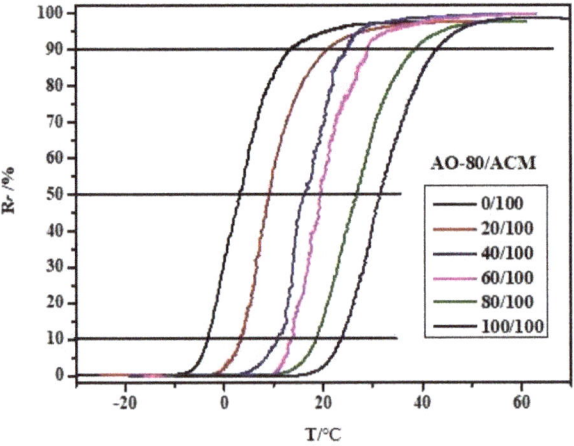

Figure 9. R_r–T curves of AO-80/ACM rubber composites.

Figure 10 shows the shape-memory recovery of AO-80/ACM (100/100) rubber composite. After placing the components in water at 20 °C, which is higher than T_g, they gradually recovered their original shape (Figure 10, t = 9 s–5 min). The results indicate that AO-80/ACM rubber composites exert shape-memory effects.

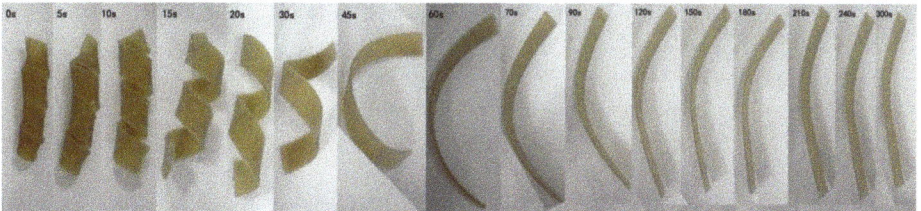

Figure 10. Shape recovery of AO-80/ACM rubber composites from a spiral-shaped temporary shape to stretched strip in water at 20 °C, which is higher than T_g.

3.5. Dynamic Mechanical Properties of AO-80/ACM Rubber Composites

Dynamic mechanical properties of AO-80/ACM rubber composites are shown in Figure 11. All curves have only one transition, and the curves moved toward higher temperatures with an increasing dosage of AO-80. The E' values of the AO-80/ACM rubber composites were similar in the glassy regions, whereas the E' values in the rubbery regions decreased with an increasing dosage of AO-80. This was because the E' values of AO-80 were similar to that of ACM matrix; therefore the E' values of AO-80/ACM rubber composites were similar in the glassy state. When AO-80/ACM rubber composites were in the rubbery state, temperature was higher than the T_g of AO-80 (40.9 C) [44], the AO-80 acted as a plasticizer after becoming soft, therefore the E' values of AO-80/ACM rubber composites decreased. In AO-80/ACM rubber composites, all specimens showed a difference of approximately three orders of magnitude of AO-80/ACM rubber composites, which is responsible for the good recovery ratio and good shape fixity ratio for all specimens.

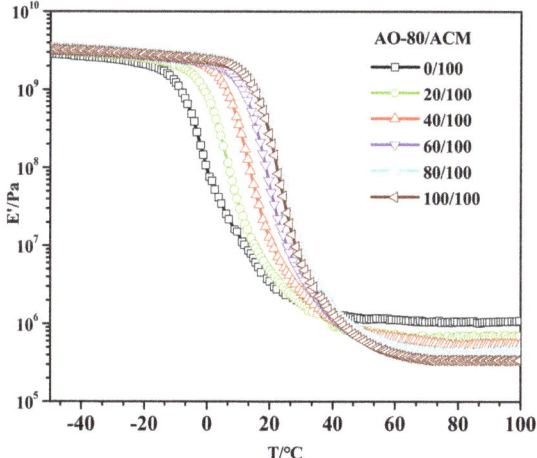

Figure 11. E'–T curves of AO-80/ACM rubber composites.

4. Conclusions

In this work, AO-80/ACM rubber composites were prepared. AO-80 has been successfully used to tailor T_{trans} and T_g of AO-80/ACM rubber composites became higher with the increment in AO-80. The formation of hydrogen bonding between carbonyl and ether groups of ACM molecules and the -OH of AO-80 is responsible for the increase in T_g. Considering that T_{trans} of ACM and AO-80/ACM rubber composites was related to T_g, the T_{trans} of AO-80/ACM rubber composites shifted from −11 °C to 10 °C when the dosage of AO-80 was added from zero phr to one hundred phr. In shape-memory experiments, the composites presented a shape-memory effect, and T_{10}, T_{50}, and T_{90} increased with

T_{trans}. Shape memory can be maintained at a wide deformation range and has good repeatability. All memory tests led to the conclusion that AO-80/ACM rubber composites feature excellent shape behavior. R_f and R_r of AO-80/ACM rubber composites were higher than 99% and 99%, respectively. The aforementioned approaches of tuning the transition temperature of developed composites can be potentially applied to other polymer systems.

Author Contributions: X.-y.Z. conceived and designed the experiments; S.-k.H. and S.C. performed the experiments; L.-q.Z. analyzed the data; M.-m.G. contributed reagents/materials/analysis tools; S.-k.H. wrote the paper.

Funding: This research was funded by [National Natural Science Foundation of China] grant number [51103006 and 51320105012].

Acknowledgments: We are thankful to Bao-chun Guo from the South China University of Technology and Wei-yu Cao from Beijing University of Chemical Technology for measurement of shape memory properties.

Conflicts of Interest: The authors declare no conflict of interest.

References

1. Weiss, R.A.; Izzo, E.; Mandelbaum, S. New Design of Shape Memory Polymers: Mixtures of an Elastomeric Ionomer and Low Molar Mass Fatty Acids and Their Salts. *Macromolecules* **2008**, *41*, 2978–2980. [CrossRef]
2. Lendlein, A.; Kelch, S. Shape-Memory Polymers. *Angew. Chem. Int. Ed.* **2002**, *41*, 2034–2057. [CrossRef]
3. Sun, L.; Huang, W.M.; Ding, Z.; Zhao, Y.; Wang, C.C.; Purnawali, H.; Tang, C. Stimulus-responsive shape memory materials: A review. *Mater. Des.* **2012**, *33*, 577–640. [CrossRef]
4. Leng, J.S.; Lan, X.; Liu, Y.L.; Du, S.Y. Shape-memory polymers and their composites: Stimulus methods and applications. *Prog. Mater. Sci.* **2011**, *56*, 1077–1135. [CrossRef]
5. Lendlein, A.; Jiang, H.Y.; Jünger, O.; Langer, R. Light-induced shape-memory polymers. *Nature* **2005**, *434*, 879–882. [CrossRef] [PubMed]
6. Fabrizio, Q.; Loredana, S.; Anna, S.E. Shape memory epoxy foams for space applications. *Mater. Lett.* **2012**, *69*, 20–23. [CrossRef]
7. Lendlein, A.; Behl, M.; Hiebl, B.; Wischke, C. Shape-memory polymers as a technology platform for biomedical applications. *Expert Rev. Med. Devices* **2010**, *7*, 357–379. [CrossRef]
8. Liu, Y.P.; Gall, K.; Dunn, M.L.; Mcculskey, P. Thermomechanics of shape memory polymer nanocomposites. *Mech. Mater.* **2004**, *36*, 929–940. [CrossRef]
9. Squeo, E.A.; Quadrini, F. Shape memory epoxy foams by solid-state foaming. *Smart Mater. Struct.* **2010**, *19*, 533–536. [CrossRef]
10. Liu, Y.Y.; Han, C.M.; Tan, H.F.; Du, X.W. Thermal, mechanical and shape memory properties of shape memory epoxy resin. *Mater. Sci. Eng. A* **2010**, *527*, 2510–2514. [CrossRef]
11. Yang, B.; Huang, W.M.; Li, C.; Li, L. Effects of moisture on the thermomechanical properties of a polyurethane shape memory polymer. *Polymer* **2006**, *47*, 1348–1356. [CrossRef]
12. Huang, W.M.; Yang, B.; An, L.; Li, C. Water-driven programmable polyurethane shape memory polymer: Demonstration and mechanism. *Appl. Phys. Lett.* **2005**, *86*, 114105–114108. [CrossRef]
13. Lee, K.M.; Koerner, H.; Vaia, R.A.; Bunning, T.J.; White, T.J. Light-activated shape memory of glassy, Azobenzene liquid crystalline polymer networks. *Soft Matter* **2011**, *7*, 4318–4324. [CrossRef]
14. Koerner, H.; Price, G.; Pearce, N.A.; Alxander, M.; Vaia, R.A. Remotely actuated polymer nanocomposites-stress-recovery of carbon-nanotube-filled thermoplastic elastomers. *Nat. Mater.* **2004**, *3*, 115. [CrossRef] [PubMed]
15. Xiao, Y.Y.; Gong, X.L.; Kang, Y.; Jiang, Z.C.; Zhang, S.; Li, B.J. Light-, pH- and thermal-responsive hydrogels with the triple-shape memory effect. *Chem. Commun.* **2016**, *52*, 10609–10612. [CrossRef]
16. Yang, J.; Wen, H.; Zhuo, H.; Chen, S.; Ban, J. A new type of photo-thermo staged-responsive shape-memory polyurethanes network. *Polymers* **2017**, *9*, 287–297. [CrossRef]
17. Ji, F.L.; Zhu, Y.; Hu, J.L.; Liu, Y.; Yeung, L.Y.; Ye, G.D. Smart polymer fibers with shape memory effect. *Smart Mater. Struct.* **2006**, *15*, 1547. [CrossRef]

18. Du, F.P.; Ye, E.Z.; Yang, W.; Shen, T.H.; Tang, C.Y.; Xie, X.L.; Zhou, X.P.; Law, W.C. Electroactive shape memory polymer based on optimized multi-walled carbon nanotubes/polyvinyl alcohol nanocomposites. *Compos. Part B* **2015**, *68*, 170–175. [CrossRef]
19. Lu, H.B.; Liu, Y.J.; Gou, J.H.; Leng, J.S.; Du, S.Y. Synergistic effect of carbon nanofiber and carbon nanopaper on shape memory polymer composite. *Appl. Phys. Lett.* **2010**, *96*, 879. [CrossRef]
20. Lu, H.B.; Liu, Y.J.; Gou, J.H.; Leng, J.S.; Du, S.Y. Synergistic effect of carbon nanofiber and sub-micro filamentary nickel nanostrand on the shape memory polymer nanocomposite. *Smart Mater. Struct.* **2011**, *20*, 035017–035023. [CrossRef]
21. Lendlein, A.; Langer, R. Biodegradable, elastic shape-memory polymers for potential biomedical applications. *Science* **2002**, *296*, 1673–1676. [CrossRef] [PubMed]
22. Han, X.J.; Dong, Z.Q.; Fan, M.M.; Liu, Y.; Li, J.H.; Wang, Y.F.; Yuan, Q.J. pH-induced shape-memory polymers. *Macromol. Rapid Commun.* **2012**, *33*, 1055–1060. [CrossRef] [PubMed]
23. Meng, H.; Xiao, P.; Gu, J.; Wen, X.; Xu, J.; Zhao, C.; Zhang, J.; Chen, T. Self-healable macro-/microscopic shape memory hydrogels based on supramolecular interactions. *Chem. Commun.* **2014**, *50*, 12277–12280. [CrossRef] [PubMed]
24. Meng, H.; Zheng, J.; Wen, X.; Cai, Z.; Zhang, J.; Chen, T. Ph- and sugar-induced shape memory hydrogel based on reversible phenylboronic acid–diol ester bonds. *Macromol. Rapid Commun.* **2015**, *36*, 533–537. [CrossRef] [PubMed]
25. Yasin, A.; Li, H.Z.; Lu, Z.; Rehman, S.; Siddig, M.; Yang, H.Y. A shape memory hydrogel induced by the interactions between metal ions and phosphate. *Soft Matter* **2014**, *10*, 972. [CrossRef]
26. Ahn, S.; Deshmukh, P.; Kasi, R.M. Shape Memory Behavior of Side-Chain Liquid Crystalline Polymer Networks Triggered by Dual Transition Temperatures. *Macromolecules* **2010**, *43*, 7330–7340. [CrossRef]
27. Liu, C.D.; Chun, S.B.; Mather, P.T.; Zhang, L.; Haley, E.H.; Coughlin, E.B. Chemically Cross-Linked Polycyclooctene: Synthesis, Characterization, and Shape Memory Behavior. *Macromolecules* **2002**, *35*, 9868–9874. [CrossRef]
28. Liu, G.; Ding, X.; Cao, Y.; Zheng, Z.H.; Peng, Y.X. Shape Memory of Hydrogen-Bonded Polymer Network/Poly(ethylene glycol) Complexes. *Macromolecules* **2014**, *37*, 2228–2232. [CrossRef]
29. Cao, Y.P.; Guan, Y.; Du, J.; Luo, J.; Peng, Y.X.; Chan, A.S.C. Hydrogen-bonded polymer network-poly(ethylene glycol) complexes with shape memory effect. *J. Mater. Chem.* **2002**, *12*, 2957–2960. [CrossRef]
30. Liu, C.; Qin, H.; Mather, P.T. Review of progress in shape-memory polymers. *J. Mater. Chem.* **2007**, *17*, 1543–1558. [CrossRef]
31. Li, J.; Viveros, J.A.; Wrue, M.H.; Anthamatten, M. Shape-memory effects in polymer networks containing reversibly associating side-groups. *Adv. Mater.* **2007**, *19*, 2851–2855. [CrossRef]
32. Ware, T.; Hearon, K.; Lonnecker, A.; Wooley, K.L.; Maitland, D.J.; Voit, W. Triple-Shape Memory Polymers Based on Self-Complementary Hydrogen Bonding. *Macromolecules* **2012**, *45*, 1062. [CrossRef] [PubMed]
33. Chen, L.; Li, W.B.; Liu, Y.J.; Leng, J.S. Nanocomposites of epoxy-based shape memory polymer and thermally reduced graphite oxide: Mechanical, thermal and shape memory characterizations. *Compos. Part B* **2016**, *91*, 75–82. [CrossRef]
34. Lendlein, A.; Behl, M. Shape-memory polymers. *Mater. Today* **2007**, *10*, 20–28. [CrossRef]
35. Zhao, X.Y.; Cao, Y.J.; Zou, H.; Li, J.; Zhang, L.Q. Structure and Dynamic Properties of Nitrile- Butadiene Rubber /Hindered Phenol Composites. *J. Appl. Polym. Sci.* **2011**, *123*, 3696–3702. [CrossRef]
36. Zhao, X.Y.; Xiang, P.; Cao, Y.J.; Tian, M.; Fond, H.; Jin, R.G.; Zhang, L.Q. Nitrile butadiene rubber/hindered phenol nanocomposites with improved strength and high damping performance. *Polymer* **2007**, *48*, 6056–6063. [CrossRef]
37. Jiang, H.Y.; Kelch, S.; Lendlein, A. Polymers Move in Response to Light. *Adv. Mater.* **2006**, *18*, 1471–1475. [CrossRef]
38. Parameswaranpillai, J.; Ramanan, S.P.; Jose, S.; Siengchin, S.; Magueresse, A.; Janke, A.; Pionteck, J. Shape memory properties of epoxy/PPO-PEO-PPO triblock copolymer blends with tunable thermal transitions and mechanical characteristics. *Ind. Eng. Chem. Res.* **2017**, *56*, 14069–14077. [CrossRef]
39. Kumar, K.S.S.; Biju, R.; Nair, C.P.R. Progress in shape memory epoxy resins. *React. Funct. Polym.* **2013**, *73*, 421–430. [CrossRef]

40. Yu, R.; Yang, X.; Zhang, Y.; Zhao, X.; Wu, X.; Zhao, T.; Zhao, Y.; Huang, W. Three-dimensional printing of shape memory composites with epoxy-acrylate hybrid photopolymer. *ACS Appl. Mater. Interfaces* **2017**, *9*, 1820–1829. [CrossRef]
41. Wang, W.; Liu, D.; Liu, Y.; Leng, J.; Bhattacharyya, D. Electrical actuation properties of reduced graphene oxide paper/epoxy-based shape memory composites. *Compos. Sci. Technol.* **2015**, *106*, 20–24. [CrossRef]
42. Karger-Kocsis, J.; Kéki, S. Review of Progress in Shape Memory Epoxies and Their Composites. *Polymers* **2018**, *10*, 34. [CrossRef]
43. Xiao, D.L.; Zhao, X.Y.; Feng, Y.P.; Xiang, P.; Zhang, L.Q.; Wang, W.M. The structure and dynamic properties of thermoplastic polyurethane elastomer/hindered phenol hybrids. *J. Appl. Polym. Sci.* **2010**, *116*, 2143–2150. [CrossRef]
44. Ghobadi, E.; Heuchel, M.; Kratz, K.; Lendlein, A. Atomistic Simulation of the Shape-Memory Effect in Dry and Water Swollen Poly[(rac-lactide)-*co*-glycolide] and Copolyester Urethanes Thereof. *Macromol. Chem. Phys.* **2014**, *215*, 65–75. [CrossRef]
45. Ghobadi, E.; Heuchel, M.; Kratz, K.; Lendlein, A. Simulation of volumetric swelling of degradable poly[(rac-lactide)-*co*-glycolide] based polyesterurethanes containing different urethane-linkers. *J. Appl. Biomater. Funct. Mater.* **2013**, *10*, 293–301. [CrossRef] [PubMed]
46. Cao, Y.Y.; Mou, H.Y.; Shen, F.; Xu, H.Y.; Hu, G.H.; Wu, C.F. Hydrogenated nitrile butadiene rubber and hindered phenol composite. II. Characterization of hydrogen bonding. *Polym. Eng. Sci.* **2011**, *51*, 201–208. [CrossRef]
47. Wu, C.F. Microstructural development of a vitrified hindered phenol compound during thermal annealing. *Polymer* **2003**, *44*, 1697–1703. [CrossRef]
48. Zhao, X.Y.; Lu, Y.L.; Xiao, D.L.; Wu, S.Z.; Zhang, L.Q. Thermoplastic Ternary Hybrids of Polyurethane, Hindered Phenol and Hindered Amine with Selective Two-Phase Dispersion. *Macromol. Mater. Eng.* **2009**, *294*, 345–351. [CrossRef]

 © 2018 by the authors. Licensee MDPI, Basel, Switzerland. This article is an open access article distributed under the terms and conditions of the Creative Commons Attribution (CC BY) license (http://creativecommons.org/licenses/by/4.0/).

Article

Control-Oriented Modelling of a 3D-Printed Soft Actuator

Ali Zolfagharian [1,*], Akif Kaynak [1,*], Sui Yang Khoo [1], Jun Zhang [1], Saeid Nahavandi [2] and Abbas Kouzani [1,*]

1   School of Engineering, Deakin University, 3216 Geelong, Australia; sui.khoo@deakin.edu.au (S.Y.K.); tzy@deakin.edu.au (J.Z.)
2   Institute for Intelligent Systems Research and Innovation (IISRI), Deakin University, 3216 Geelong, Australia; saeid.nahavandi@deakin.edu
*   Correspondence: a.zolfagharian@deakin.edu.au (A.Z.); akaynak@deakin.edu.au (A.K.); abbas.kouzani@deakin.edu (A.K.)

Received: 15 November 2018; Accepted: 24 December 2018; Published: 26 December 2018

Abstract: A new type of soft actuator was developed by using hydrogel materials and three-dimensional (3D) printing technology, attracting the attention of researchers in the soft robotics field. Due to parametric uncertainties of such actuators, which originate in both a custom design nature of 3D printing as well as time and voltage variant characteristics of polyelectrolyte actuators, a sophisticated model to estimate their behaviour is required. This paper presents a practical modeling approach for the deflection of a 3D printed soft actuator. The suggested model is composed of electrical and mechanical dynamic models while the earlier version describes the actuator as a resistive-capacitive (RC) circuit. The latter model relates the ionic charges to the bending of an actuator. The experimental results were acquired to estimate the transfer function parameters of the developed model incorporating Takagi-Sugeno (T-S) fuzzy sets. The proposed model was successful in estimating the end-point trajectory of the actuator, especially in response to a broad range of input voltage variation. With some modifications in the electromechanical aspects of the model, the proposed modelling method can be used with other 3D printed soft actuators.

Keywords: modeling; soft actuator; soft robot; 3D print

---

1. Introduction

In contrast to building robots from rigid materials in conventional robotics, soft, responsive, flexible, and compliant materials have been used to make composite materials that can mimic biological systems. The study of such responsive composites is often referred to as soft robotics, which is a specific category of the field of robotics. There has been a growing interest in these materials, which have potential applications in sensors and actuators. Functional components composed of soft and active materials can be 3D-printed. Furthermore, it was observed that monolithic structures independent of pneumatics or fluidics systems, which have their own shortcomings [1], can be developed as muscle-like actuators by means of materials such as shape memory alloys, shape memory polymers, or responsive hydrogel materials [1–3].

Using 3D bio-printing as part of the fabrication of soft actuators from responsive materials such as hydrogel actuators with multi-material compositions that have spatial control was recently investigated [4]. In a recent study, polyelectrolyte chitosan hydrogel was used in the 3D-printing of a soft actuator [5]. Chitosan was found to be an appropriate material for drug release, cell manipulation, and similar biological applications where the actuation and responsiveness to external stimuli is essential [6]. This is mainly referred to its antibacterial properties, which are attributed to free amino groups on the hydrogel backbone.

From one point of view, hydrogels can be divided into two groups in terms of responsive behaviour to the potential gradient. A group known as non-ionic hydrogels do not demonstrate a notable actuation response to input voltage when immersed in electrolyte solutions due to even distribution of hydrated ions on two sides of hydrogels. Polyelectrolyte hydrogels, however, react differently from non-ionic hydrogels when immersed into electrolyte solutions. Upon immersion in electrolyte solutions and applying voltages, polyelectrolyte hydrogels show actuation behaviour based on parameters like strength and polarity of input voltages as well as the pH of solutions [7]. Various key factors, such as ionic charge and crosslinking densities as well as polymer and external electrolyte concentrations [7], have been identified to affect the amount of bending of polyelectrolyte hydrogels in response to an input voltage. For chitosan hydrogel, when is immersed into an electrolyte solution with a high pH, the carboxylic group would be deprotonated in order for the hydrogel actuator to take a negative charge. Then, upon the application of voltage potentials on electrodes, the hydrated cations enter to one side of the actuator more than the other side. This result in an ionic strength difference between inside and outside of the hydrogel leads to a greater osmotic pressure on one side of the hydrogel than the other side, which, in turn, leads to a bending of the hydrogel actuator to the counter electrode or cathode. Thus, the gel near the anode swells and causes it to bend toward the cathode [8]. Along with the osmotic pressure gradient, the ionic strength also affects the extent of bending of the polyelectrolyte soft actuator. Due to a higher electric voltage, the movements of ions are accelerated, which lead to faster ionic and osmotic pressure gradients and, therefore, increase both the bending angle and the bending rate of the actuator. Yet, the relation of bending behaviour of the actuator is not always proportional to ionic strength of electrolyte solution because of the shielding effect phenomenon [9].

Polyelectrolyte soft actuators have been considered as electro-chemo-mechanical systems that make their modelling quite complicated. The bending and actuation performances of polyelectrolyte soft actuators are influenced by uncertainties and time-varying parameters stemming from back relaxation and modulus variation phenomena. These undesirable factors should be dealt with for further fabrication and applications of such soft actuators. However, with the incorporation of 3D-printing in manufacturing such soft actuators, developing a control-oriented model for estimating the behaviour of these 3D-printed polyelectrolyte soft actuators in real world applications is in high demand.

Studies on dynamics modelling of conventional polyelectrolyte soft actuators have been conducted. Black-box models were used in some works for calculating the curvature of the actuator based on input voltage but were not scalable and too simple to describe the actuator performance entirely [10]. Advanced grey-box models based on electrical circuit models, like RC [11] and distributed transmission line models [12], have also been developed to correlate the applied voltage to the bending of the actuators [13]. More complex white-box models [10–12] considering complicated electro-chemo-mechanical principals were developed to explain the more insightful details of underlying physics for accurate dynamics modelling of the polyelectrolyte actuators. Yet, these models were too complex and not appropriate for real time control application of such actuators. This study establishes a mathematical grey-box relation of the 3D-printed polyelectrolyte soft actuator by coupling both mechanical and electrical dynamics of the actuators.

Takagi-Sugeno (T-S) fuzzy modelling has been a practical approach in control designs applications [14,15]. This study develops a reliable model of the 3D-printed polyelectrolyte soft actuator based on the T-S fuzzy modelling strategy. The proposed model relates the different input voltages applied to the actuator to the bending of the actuator via a universal T-S model surfing among the voltage dependent sub-models. This provides a scalable and practical model for further control applications of such systems.

The rest of the paper is comprised of the following sections. First, 3D printing of the polyelectrolyte soft actuator using chitosan is explained. Then, an electro-chemo-mechanical model of the 3D-printed polyelectrolyte actuator is developed. Lastly, the developed model is validated via the experimental test data.

2. Fabrication of the Polyelectrolyte Actuator

Like other 3D-printed products, first, the actuator model was drawn in Solidworks 2016 (Dassault Systemes, Waltham, MA, USA) and then the model was imported into an EnvisionTEC GmbH Bioplotter V2.2 software program (EnvisionTEC, Gladbeck, Germany). The required materials including medium molecular weight chitosan (with 75–85% deacetylation degree) and acetic acid solution were purchased from SigmaAldrich (Sydney, NSW, Australia). A mixture of 1.6 g chitosan in 0.8 mL acetic acid solution (1 v/v%) was made under vigorous stirring at 50 °C for 2 h. The resultant 3D-print ink was then prepared through sonication and centrifugation. Lastly, the ready-to-print ink was poured into a low-temperature 3D Bioplotter syringe. For solidifying each layer after extrusion, a solution of Ethanolic Sodium Hydroxide (EtOH-NaOH) with 0.25 M NaOH (Sigma Aldrich), 70 v/v% EtOH (3:7 ratio), was prepared. The porous chitosan beam with the size of 40 mm by 8 mm by 2 mm was printed layer-by-layer (Figure 1a). The 3D printing was performed with optimized parameters of the 3D Bioplotter, as explained extensively in an earlier work by the authors [16].

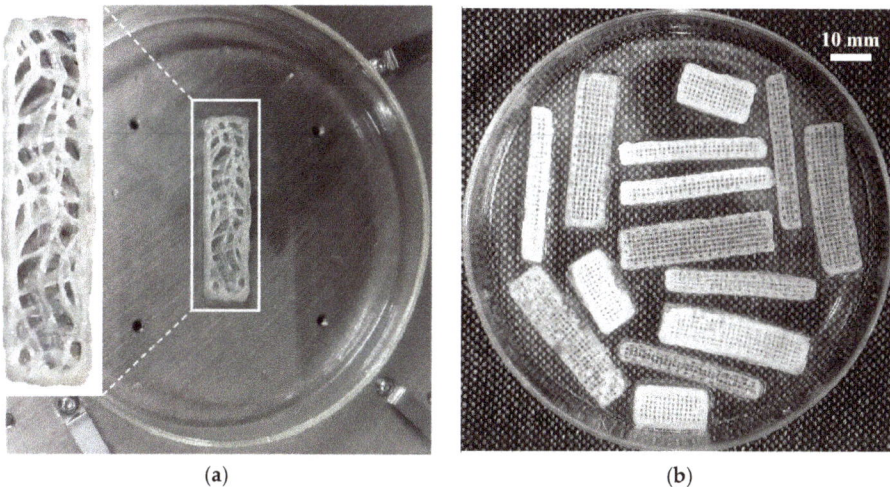

Figure 1. (a) An arbitrary pattern 3D printed polyelectrolyte actuator and (b) different patterns and sizes of 3D-printed polyelectrolyte actuators.

3. Electro-Chemo-Mechanical Model of the 3D-Printed Polyelectrolyte Actuator

To circumvent the complexity of multiphysics modeling, this study suggests a scalable model of the polyelectrolyte actuator. Doing so, various factors are considered in the behavior of the actuators to estimate the dynamics parameters more realistically. Thus, the actuator model comprises both electrochemical and electromechanical models considering their dynamics coupling.

3.1. Electrochemical Modelling

Figure 2 shows an electrochemical RC model that gives the relation between applied voltage and current flow across the polyelectrolyte gel. A diffusion model to represent the current flow across the 3D-printed polyelectrolyte actuator is calculated from the Fick's law of diffusion as:

$$i_D(t) = -F \times A \times D \times \frac{\partial c(y,t)}{\partial y}, \tag{1}$$

where c refers to the ion concentration, y indicates the path along the thickness of the actuator, D is the diffusion coefficient constant, A is the area between the interface of electrolyte solution and polyelectrolyte actuator, and F denotes the Faraday constant. Then, the current running across the double layer capacitance can be calculated assuming the double-layer capacitance thickness as δ:

$$i_C(t) = F \times A \times \delta \times \frac{\partial c(y,t)}{\partial y}. \quad (2)$$

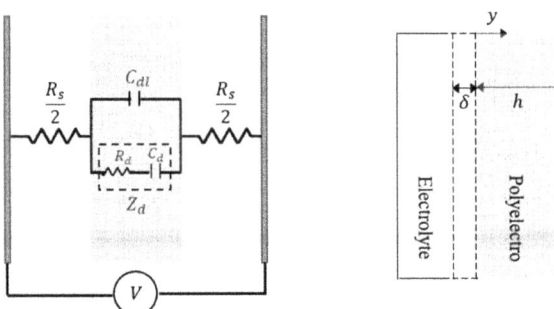

Figure 2. Equivalent RC circuit where Z_d is the diffusion impedance, C_{dl} denotes the double layer capacitance, and R_s refers to the resistance of the electrolyte solution.

Next, for a 3D-printed polyelectrolyte actuator with the thickness of the h, the diffusion equation model can be written as:

$$\frac{\partial c(y,t)}{\partial t} = D \frac{\partial^2 c(y,t)}{\partial y^2} \quad 0 < y < h, \quad \frac{\partial c(y,t)}{\partial t}(@y = h) = 0. \quad (3)$$

Lastly, an electrochemical model relating the current flow across the actuator to the applied voltage can be obtained by simultaneously solving Equations (1)–(3) as:

$$\frac{I(s)}{V(s)} = \frac{s\left[\frac{\sqrt{D}}{\delta}\tanh\left(h\sqrt{\frac{s}{D}}\right) + \sqrt{s}\right]}{\frac{\sqrt{s}}{C_{dl}} + s\sqrt{s}R_s + R_s\frac{\sqrt{D}}{\delta}\stanh\left(h\sqrt{\frac{s}{D}}\right)}. \quad (4)$$

3.2. Electromechanical Modelling

An electromechanical model that relates the osmotic pressure caused by input voltage to the bending of 3D-printed polyelectrolyte actuator is developed in this section. Assuming the strain-to-charge ratio α, the relation between the charges density ρ_{ch} and the in-plane strain ε in polyelectrolyte actuators can be written as [17,18]:

$$\varepsilon = \alpha \rho_{ch}, \quad (5)$$

where, for a 3D-printed soft actuator with a size of W and L as width and length, respectively, ρ_{ch} can be calculated as:

$$\rho_{ch} = \frac{1}{WLh} \int_0^t I(t)dt \xrightarrow{L} \rho_{ch}(s) = \frac{I(s)}{sWLh}, \quad (6)$$

Generally, actuators strains are calculated knowing the tip displacement as well as the initial length and thickness of the actuators when the bending curvature is comparatively small. However, the curvature radius is not negligible when the soft actuators experience larger bending. In this study, as shown in Figure 3 and Equation (7), the strain ε resulted purely from the bending of the actuator, which is obtained as:

$$\varepsilon = \frac{l_0 - l_c}{l_c} = \frac{(R+y)\theta - R\theta}{R\theta} = \frac{y_b}{R} = Ky_b, \quad (7)$$

where l_o and l_c are the swelled length and the centered length of actuators, θ and R refer to the bending angle and the radius of curvature of the actuator, respectively, y_b shows the distance between the outer edge of the swelled layer and the reference plane, and K is the bending curvature of the actuator. To estimate an accurate value of the stress induced on the bending of the actuator, the superposition of the actuations effect and the bending strain are considered together. Therefore, the total induced stress, assuming the Young's modulus of the 3D printed polyelectrolyte actuator as E, can be calculated as follows.

$$\sigma = E(K \pm \alpha \rho_{ch}), \tag{8}$$

where the opposite signs represent the expansion and contraction of the actuator sides.

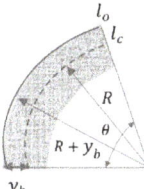

Figure 3. Schematic of the bending of the printed actuator.

Making the total moments and forces on the actuators equal zero, the bending curvature K can be realized. This is somewhat related to a new defined coefficient φ as follows.

$$\sum F = \int_{-h/2}^{0} E(Ky_b + \alpha\rho_{ch})y_b dy_b + \int_{0}^{h/2} E(Ky_b - \alpha\rho_{ch})y_b dy_b = 0,$$
$$\sum \sigma = \int_{-h/2}^{0} E(Ky_b + \alpha\rho_{ch})dy_b + \int_{0}^{h/2} E(Ky_b - \alpha\rho_{ch})dy_b = 0, \tag{9}$$
$$K = \varphi \rho_{ch}, \quad \text{where } \varphi = \frac{3\alpha}{Eh}.$$

Combining Equations (9) and (6) into Equation (4), the bending curvature can be expressed as:

$$K(s) = \frac{\varphi V(s)}{WLh} \frac{\left[\frac{\sqrt{D}}{\delta}\tanh\left(h\sqrt{\frac{s}{D}}\right) + \sqrt{s}\right]}{\frac{\sqrt{s}}{C_{dl}} + R_{ss}\sqrt{s} + R_s \frac{\sqrt{D}}{\delta} s\tanh\left(h\sqrt{\frac{s}{D}}\right)}. \tag{10}$$

Lastly, the model relating the applied voltage to the bending of the 3D-printed polyelectrolyte actuator can be obtained by incorporating Equation (6) and Equations (5)–(10) in the following equation [19].

$$\frac{Y(s)}{V(s)} = \frac{\varphi}{WLh} \frac{1}{sR_s + \frac{1}{C_{dl}\left[1 + \frac{\sqrt{D}}{\delta\sqrt{s}}\tanh\left(h\sqrt{\frac{s}{D}}\right)\right]}}, \quad \text{where, } Y(s) = \frac{1}{K(s)} \pm \sqrt{\frac{1}{K(s)^2} - L^2}. \tag{11}$$

Equation (11) needs to be restructured to be practical for control applications due to the presence of a hyperbolic tangent term in its denominator, which is dimensionally infinite. To deal with this, the dimensionally infinite hyperbolic tangent term can be replaced by Mittag Leffler's expansion as:

$$\frac{\tanh\left(\frac{1}{2}\sqrt{\frac{X}{Y}}\right)}{4\sqrt{XY}} = \sum_{n=0}^{\infty} \frac{1}{X + n^2((2n+1)^2 Y)}, \tag{12}$$

where $X = s$ and $Y = D/4h^2$. This simplification confines the model in a bounded range of frequencies for input voltages within the low interest frequency range for the 3D-printed polyelectrolyte soft actuator. Hence, Equation (12) represents the model for the small values of n, so a fourth order model approximates bending displacement of the 3D-printed soft actuator as:

$$\frac{Y(s)}{V(s)} = e^{\beta s} \frac{(s+z_1)(s+z_2)(s+z_3)}{(s+p_1)(s+p_2)(s+p_3)(s+p_4)}, \tag{13}$$

where β is a real value and z_i's (n = 1, 2, 3) are the zeros and p_i's (n = 1, 2, 3, 4) are the poles of the soft actuator transfer function.

3.3. T-S Fuzzy System Modeling Formulation

Consider a simplified dynamic system without uncertainty systems as:

$$\dot{x} = A(x) + B(x)u. \tag{14}$$

Fuzzy inference rules for the T-S fuzzy dynamic model of Equation 16 can be described as follows [20].

R^i : IF z_1 is w_1^i AND ... z_n is w_n^i

THEN

$$\dot{x} = A^i(x) + B^i(x)u \text{ for } i = 1, \ldots, m, \tag{15}$$

where the matrices $A^i \in R^{n \times n}$ and $B^i \in R^{n \times 1}$ represent the subsystem parameters, $z = (z_1, \ldots, z_n)^T$, behaviour. x is the system state variable vector and u is the system input. R^i represents the ith fuzzy inference rule, $z(t) = [z_1(t)\ z_2(t) \ldots z_n(t)]$ among m number of inference rules.

Knowing $M_j (j = 1, \ldots, n)$ are the fuzzy sets, the inferred fuzzy set w^i can be used to calculate μ_i as the normalized fuzzy membership function of inferred fuzzy sets as follows.

$$w^i = \prod_{j=1}^{n} M_j^i, \quad \mu^i = \frac{w^i}{\sum_{i=1}^{m} w^i}, \text{ and } \sum_{i=1}^{m} \mu^i = 1. \tag{16}$$

The singleton fuzzifier, product inference, and center-average defuzzifier are used to form the global dynamic fuzzy model of the nominal system of Equation (17) as:

$$\dot{x} = A(\mu)x + B(\mu)u(t),$$
$$A(\mu) = \sum_{i=1}^{m} \mu^i A^i, \ B(\mu) = \sum_{i=1}^{m} \mu^i B^i, \text{ where } \mu = (\mu^1, \mu^2, \ldots, \mu^m). \tag{17}$$

3.4. Actuator Characterization

The bending of the 3D-printed polyelectrolyte hydrogel actuator can be justified by the Donnan effect phenomenon. This means that the motion of counter-ions initiated by applied voltage leads to the ionic gradient within the hydrogel networks along the direction of the electric field. This results in an osmotic pressure difference within the hydrogel structure, and, consequently, causes the deflection of a 3D-printed actuator toward the counter electrode. Several factors can be considered to characterize the behaviour of the actuator. First, the effects of the electrolyte solution and its concentration on the bending behaviour of the 3D-printed polyelectrolyte actuator should be defined. Doing so, two different electrolyte solutions, NaOH and NaCl, were used to test the 3D-printed actuator endpoint deflection behaviour. The maximum endpoint deflection for the same sizes (40 mm × 8 mm × 2 mm) and patterns of 3D-printed actuators are measured with respect to the ionic strength of the electrolyte solutions. A constant voltage of 5 V was applied between the electrodes. The concentration of NaOH and NaCl solutions ranged from 0.1 to 0.2 M with an increment of 0.2 M for each experiment. The experiments repeated three times and the results are depicted in Figure 4a. Regardless of the concentration of electrolyte solution, the actuator reached a higher deflection in the NaOH solution than in the NaCl solution. Additionally, it is observed that, for both electrolyte solutions, there were an optimum value for electrolyte ionic strength to achieve a maximum bending angle. As shown in Figure 4, this optimum value for the 3D-printed actuator tested here was near 0.12 M. From 0.1 to 0.12 M, the endpoint deflection of the 3D-printed actuator increased. This can be attributed to an increase of the free ions moving from the surrounding solution toward their counter-electrodes. However, if the concentration of the solutions were greater than the critical concentration, 0.12 M, the shielding effect of the poly-ions leads to a reduction in the electrostatic repulsion of the poly-ions, and the subsequent decrease in the endpoint deflection. Hence, in the rest of the paper, all tests are performed in NaOH electrolyte solution of 0.12 M.

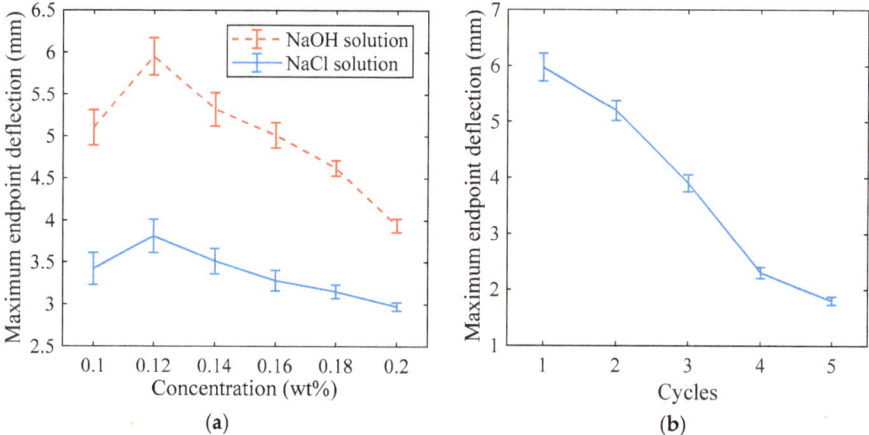

Figure 4. (a) Response of the actuators with the same pattern and dimensions to different electrolyte solutions. (b) Maximum endpoint deflections of the actuators to 5 V input signal over different cycles. Error bars indicate the standard deviation of the maximum distance measured over the three trials.

Furthermore, several different patterns and dimensions of the actuators with the same material are 3D-printed (Figure 1b). Then, the behaviour of the 3D-printed actuators based on different dimensions are investigated and the results are inserted in Table 1. The results reveal that the maximum endpoint deflection of the actuators increase in proportion to their length. It also shows almost no correlation with the width of the actuator. In addition, the bending deflection was inversely correlated to the thickness of the actuator [4].

Table 1. The results of maximum endpoint deflection for the same pattern and different sizes of 3D-printed actuators.

Actuator	Length (mm)	Width (mm)	Thickness (mm)	Maximum Deflection (mm)
S1	40	8	2	5.92
S2	40	8	1	8.27
S3	40	4	2	5.38
S4	20	8	2	1.83

To reveal hysteresis behaviour of a 3D-printed actuator, the repeatability tests were performed. The same pattern and size of 3D-printed samples of actuators were excited with a square wave electrical stimulus. The magnitudes of the square waves were selected as 5 V with the period of 220 s and the duty cycle of 50%. The duration of the experiment was set to be 5 times the period, and the performances of the actuators were compared based on change in maximum magnitudes in each cycle, as shown in Figure 4b. The results showed that all actuators reached their maximum deflections at the first set of experiments for a specific excitation voltage.

In the next stage, the actuator's endpoint deflections were tested over time based on different input voltage magnitudes, as illustrated in Figure 5. From the results, it can be deduced that the actuation performance increased in the higher voltage. However, the occurrence of electrolyses and bubbling in electrolyte caused undesirable effects and limits the application of higher voltages in such actuators.

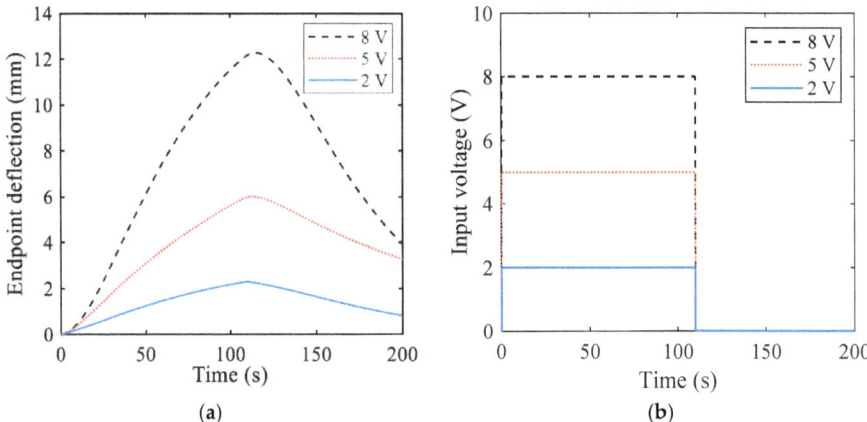

Figure 5. (a) Response of the actuators with the same pattern and dimensions to (b) different input voltages' magnitudes. The results are averaged over the three trials.

The actuator responses under various frequencies were also investigated. The square voltage of 5 V was applied between two electrodes in various frequencies including 0.0025, 0.02, 0.031, 0.054, 0.11, 0.15, and 1.1 Hz. It can be seen from Figure 6a that the maximum endpoint deflections decrease with increasing frequency since the actuators have less time to respond. In addition, response time to the first peak was measured and the results are shown in Figure 6b. From the figure, it can be observed that the actuator reaches the first peak faster as the magnitude of the peaks decreases in higher frequencies.

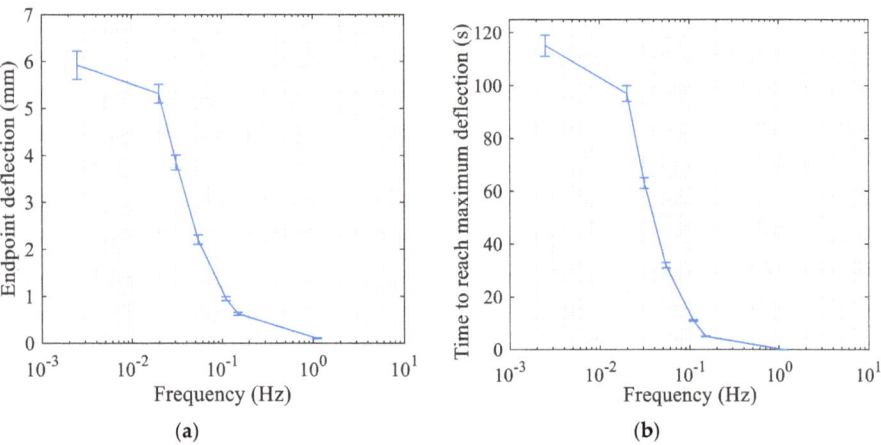

Figure 6. (a) Response of actuator's maximum endpoint deflection to various frequencies under 5 V square waves. (b) Response time to first peak. Error bars indicate the standard deviation of the maximum distance measured over the three trials.

4. Modelling and Experimental Results

The Golubev method [21] is used to estimate an appropriate model for control of the actuator [22–25]. Using the Golubev method for the input signal, the uncertain transfer function relating the applied voltage to the bending angle of the actuator is estimated as:

$$G(s) = e^{\beta s} \frac{a_1 s^3 + a_2 s^2 + a_3 s + a_4}{b_1 s^4 + b_2 s^3 + b_3 s^2 + b_4 s + b_5},\qquad(18)$$

The frequency response model of the actuators with different patterns and sizes, shown in Figure 1b, is identified based on Equation (18) and the experimental data are depicted in Figure 7. It is observed that changing various parameters of the actuator, like different sizes and patterns, lead to a new set of linear voltage dependent transfer functions. Therefore, for each specific actuator, a system identification of frequency response to different voltage levels is required. Then, the T-S fuzzy model should be incorporated to interrelate the linear transfer functions at different voltage levels for improving the model accuracy.

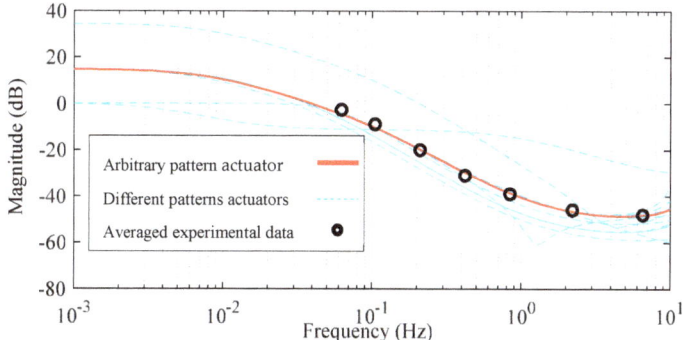

Figure 7. Frequency response system identification of different 3D-printed actuators in response to the 5V input signal.

To identify transfer function parameters for the arbitrary pattern actuator S1, lower and upper limit input signals, 2 V (shown in Figure 8a) and 8 V are applied to the system and the range of parameters identified are as follows.

$a_1 \in [0.0063, 0.0085]; a_2 \in [0.6381, 0.842]; a_3 \in [1.152, 1.924]; a_4 \in [1.14, 1.97];$
$b_1 \in [1, 1]; b_2 \in [0.3631, 0.5898]; b_3 \in [225.5, 367.2]; b_4 \in [14.98, 29.41]; b_5 \in [0.2142, 0.2275];$
$\beta = -3.67.$

Then, a two-rule based T-S model is defined as:
Rule 1: IF $|z(t)| \leq 5$ THEN $\dot{x} = A_1 x(t) + B_1 u(t)$.
Rule 2: IF $5 < |z(t)| \leq 8$ THEN $\dot{x} = A_2 x(t) + B_2 u(t)$.
Lastly, the outputs of the T-S fuzzy model (shown in Figure 7b) can be calculated as:

$$\dot{x} = \mu_1(z(t))(A_1 x(t) + B_1 u(t)) + \mu_2(z(t))(A_2 x(t) + B_2 u(t)),$$
$$\text{where } \mu_1(z(t)) + \mu_2(z(t)) = 1.\qquad(19)$$

A comparison of experimental tests with the T-S fuzzy model and estimated specific voltage models for 2 and 8 V input signals is shown in Figures 9 and 10. The data is based on the average of three experiments to confirm reproducibility. Additionally, the efficacy of the developed model in terms of scalability of the 3D-printed soft actuators with different patterns and sizes are shown in Figure 10b,c where two actuators with arbitrary and lattice patterns and sizes S1 and S4 are compared in response to an analogous input. These figures show supremacy of actuator end-point position estimation by T-S fuzzy modelling compared to specific constant voltage models when changing the 3D-printed actuator parameters such as pattern and size. Furthermore, looking at Figure 10b in detail reveals some discrepancy of the T-S model from experimental results, especially over longer experimental tests. This is attributed to the time-varying intrinsic nature of the polyelectrolyte soft actuator that demands the feedback controller for a compensation purpose in future study.

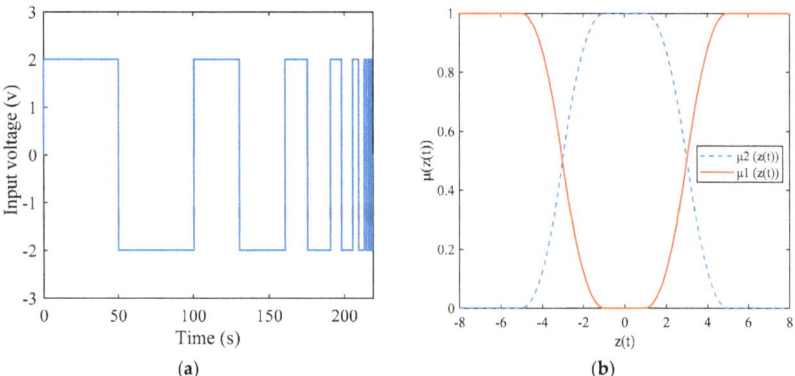

Figure 8. (a) An example of input signal for model identification (2 V). (b) Membership function of T-S model.

Figure 9. Bending of the printed actuator (a) before applying voltage; (b) under applied voltage.

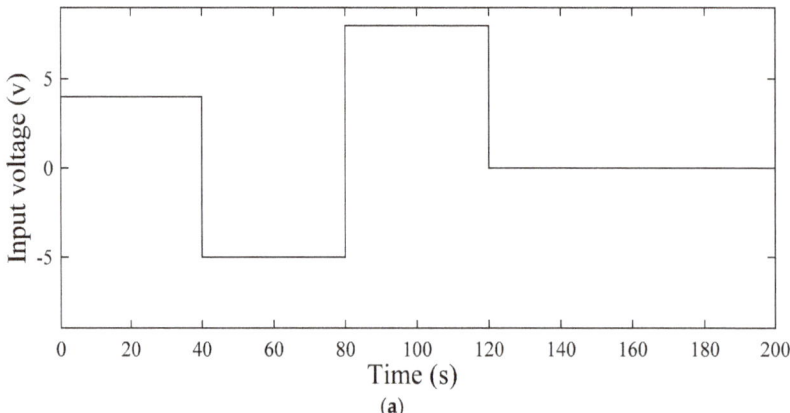

Figure 10. *Cont.*

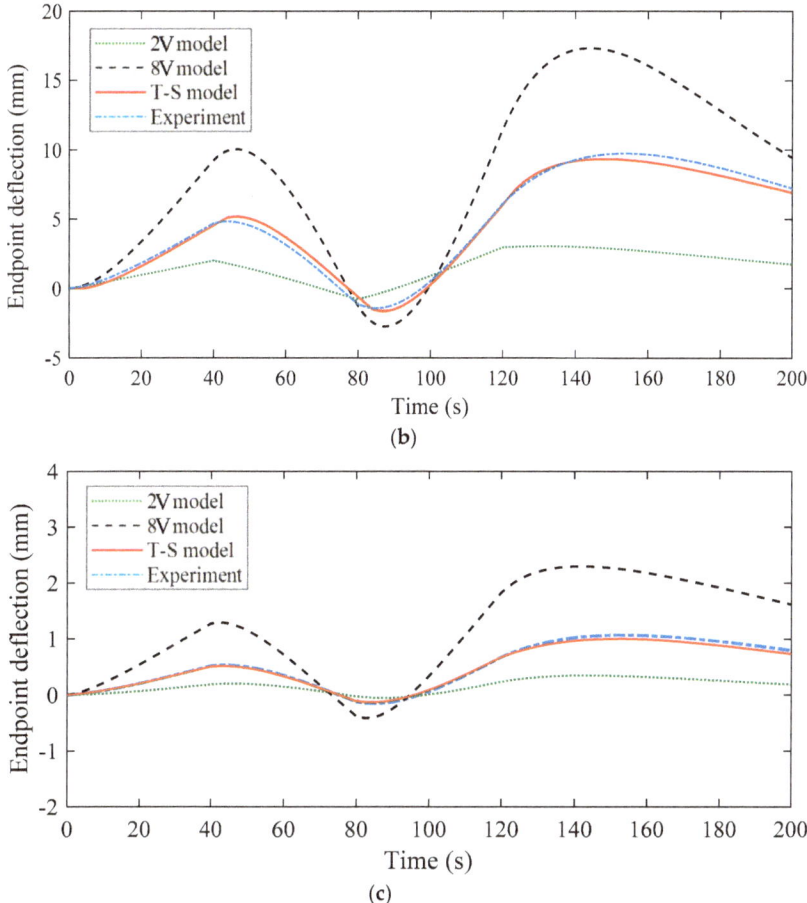

Figure 10. (a) Input voltage signal, (b) endpoint deflection of the arbitrary pattern actuator with size of S1, and (c) endpoint deflection of the lattice pattern actuator with size of S4.

5. Conclusions

A control-oriented modelling approach for 3D-printed polyelectrolyte soft actuators was presented in this study. First, a 3D-printed actuator with an arbitrary pattern was developed and characterized based on different sizes, electrolyte concentration, input magnitudes, and frequencies. Then, a linear transfer function of the 3D-printed polyelectrolyte soft actuator was developed to estimate the actuator behavior at different voltage signals. The T-S fuzzy model was further employed for a better presentation of the actuator model in a range of voltage variations by interrelating the voltage-dependent models. The experimental results showed improved performance obtained by using the T-S fuzzy model when compared to the linear transfer function at different voltages. The proposed model could be used for other 3D-printed soft actuators with custom design geometries due to its scalability.

Author Contributions: Conceptualization, A.Z.; methodology, A.Z.; software, A.Z.; validation, A.K. (Akif Kaynak) and A.K. (Abbas Kouzani); investigation, A.Z.; resources, A.K. (Abbas Kouzani); data curation, J.Z.; writing—original draft preparation, A.Z.; writing—review and editing, A.K. (Akif Kaynak) and S.N.; supervision, A.K. (Akif Kaynak), A.K. (Abbas Kouzani), and S.Y.K.; project administration, A.Z. and A.K. (Akif Kaynak); funding acquisition, A.Z.

Funding: This research received no external funding.

Conflicts of Interest: The authors declare no conflict of interest.

References

1. Zolfagharian, A.; Kouzani, A.Z.; Khoo, S.Y.; Moghadam, A.A.A.; Gibson, I.; Kaynak, A. Evolution of 3D printed soft actuators. *Sens. Actuators A Phys.* **2016**, *250*, 258–272. [CrossRef]
2. Ma, J.; Franco, B.; Tapia, G.; Karayagiz, K.; Johnson, L.; Liu, J.; Arroyave, R.; Karaman, I.; Elwany, A. Spatial control of functional response in 4D-printed active metallic structures. *Sci. Rep.* **2017**, *7*, 46707. [CrossRef] [PubMed]
3. Zolfagharian, A.; Kouzani, A.Z.; Khoo, S.Y.; Gibson, I.; Kaynak, A. 3D printed hydrogel soft actuators. In Proceedings of the IEEE Region 10—Asia and Pacific, Singapore, 22–25 November 2016; pp. 2272–2277.
4. Zolfagharian, A.; Kouzani, A.Z.; Khoo, S.Y.; Nasri-Nasrabadi, B.; Kaynak, A. Development and analysis of a 3D printed hydrogel soft actuator. *Sens. Actuators A Phys.* **2017**, *265*, 94–101. [CrossRef]
5. Zolfagharian, A.; Kaynak, A.; Khoo, S.Y.; Kouzani, A.Z. Polyelectrolyte soft actuators: 3D printed chitosan and cast gelatin. *3D Printing Addit. Manuf.* **2018**, *5*, 138–150. [CrossRef]
6. Elviri, L.; Foresti, R.; Bergonzi, C.; Zimetti, F.; Marchi, C.; Bianchera, A.; Bernini, F.; Silvestri, M.; Bettini, R. Highly defined 3D printed chitosan scaffolds featuring improved cell growth. *Biomed. Mater.* **2017**, *12*, 045009. [CrossRef] [PubMed]
7. Shiga, T.; Kurauchi, T. Deformation of polyelectrolyte gels under the influence of electric field. *J. Appl. Polym. Sci.* **1990**, *39*, 2305–2320. [CrossRef]
8. Li, Y.; Sun, Y.; Xiao, Y.; Gao, G.; Liu, S.; Zhang, J.; Fu, J. Electric field actuation of tough electroactive hydrogels cross-linked by functional triblock copolymer micelles. *ACS Appl. Mater. Interfaces* **2016**, *8*, 26326–26331. [CrossRef] [PubMed]
9. Shang, J.; Shao, Z.; Chen, X. Electrical behavior of a natural polyelectrolyte hydrogel: chitosan/carboxymethylcellulose hydrogel. *Biomacromolecules* **2008**, *9*, 1208–1213. [CrossRef] [PubMed]
10. Vunder, V.; Punning, A.; Aabloo, A. *Electromechanical Distributed Modeling of Ionic Polymer Metal Composites, Ionic Polymer Metal Composites (IPMCs)*; Shahinpoor, M., Ed.; RCS Publishing: Cambridge, UK, 2015; pp. 228–247.
11. Attaran, A.; Brummund, J.; Wallmersperger, T. Modeling and simulation of the bending behavior of electrically-stimulated cantilevered hydrogels. *Smart Mater. Struct.* **2015**, *24*, 035021. [CrossRef]
12. Bar-Cohen, Y. *Electroactive Polymer (EAP) Actuators as Artificial Muscles: Reality, Potential, and Challenges*; SPIE: Bellingham, WA, USA, 2004.
13. Liu, Y.; Zhao, R.; Ghaffari, M.; Lin, J.; Liu, S.; Cebeci, H.; de Villoria, R.G.; Montazami, R.; Wang, D.; Wardle, B.L.; et al. Equivalent circuit modeling of ionomer and ionic polymer conductive network composite actuators containing ionic liquids. *Sens. Actuators A Phys.* **2012**, *181*, 70–76. [CrossRef]
14. Tanaka, K.; Sugeno, M. Stability analysis and design of fuzzy control systems. *Fuzzy Sets Systs.* **1992**, *45*, 135–156. [CrossRef]
15. Nasiri, S.; Khosravani, M.R.; Weinberg, K. Fracture mechanics and mechanical fault detection by artificial intelligence methods: A review. *Eng. Fail. Anal.* **2017**, *81*, 270–293. [CrossRef]
16. Zolfagharian, A.; Kouzani, A.Z.; Khoo, S.Y.; Noshadi, A.; Kaynak, A. 3D printed soft parallel actuator. *Smart Mater. Struct.* **2018**, *27*, 045019. [CrossRef]
17. Madden, P.G.A. Development and Modeling of Conducting Polymer Actuators and the Fabrication of A Conducting Polymer Based Feedback Loop. Ph.D. Thesis, Massachusetts Institute of Technology, Cambridge, MA, USA, 2003.
18. Nemat-Nasser, S.; Li, J.Y. Electromechanical response of ionic polymer-metal composites. *J. Appl. Phys.* **2000**, *87*, 3321–3331. [CrossRef]
19. Fang, Y.; Tan, X.; Shen, Y.; Xi, N.; Alici, G. A scalable model for trilayer conjugated polymer actuators and its experimental validation. *Mater. Sci. Eng. C* **2008**, *28*, 421–428. [CrossRef]
20. Yu, X.; Man, Z.; Wu, B. Design of fuzzy sliding-mode control systems. *Fuzzy Sets Syst.* **1998**, *95*, 295–306. [CrossRef]
21. Golubev, B.; Horowitz, I. Plant rational transfer approximation from input-output data. *Int. J. Control* **1982**, *36*, 711–723. [CrossRef]

22. Noshadi, A.; Zolfagharian, A.; Wang, G. Performance analysis of the computed torque based active force control for a planar parallel manipulator. *Appl. Mech. Mater.* **2012**, *110*, 4932–4940. [CrossRef]
23. Zolfagharian, A.; Kouzani, A.Z.; Moghadam, A.A.A.; Khoo, S.Y.; Nahavandi, S.; Kaynak, A. Rigid elements dynamics modelling of a 3D printed soft actuator. *Smart Mater. Struct.* **2019**, *28*, 025003. [CrossRef]
24. Benedetti, I.; Nguyen, H.; Soler-Crespo, R.A.; Gao, W.; Mao, L.; Ghasemi, A.; Wen, J.; Nguyen, S.; Espinosa, H.D. Formulation and validation of a reduced order model of 2D materials exhibiting a two-phase microstructure as applied to graphene oxide. *J. Mech. Phys. Solids* **2018**, *112*, 66–88. [CrossRef]
25. Innis, P.C.; Chen, J.; Zheng, W. *Mechanism in charge transfer and electrical stability*. Conductive Polymers; CRC Press: Boca Raton, FL, USA, 2017; pp. 99–128.

© 2019 by the authors. Licensee MDPI, Basel, Switzerland. This article is an open access article distributed under the terms and conditions of the Creative Commons Attribution (CC BY) license (http://creativecommons.org/licenses/by/4.0/).

Article

Stimuli-Responsive Systems in Optical Humidity-Detection Devices

Sergio Calixto [1,*], Valeria Piazza [1] and Virginia Francisca Marañon-Ruiz [2]

1. Centro de Investigaciones en Óptica, Loma del Bosque 115, Leon C.P. 37150, Gto., Mexico; vpiazza@cio.mx
2. Laboratorio de Ciencias Químicas/Área de Química Orgánica, Departamento de Ciencias de la Tierra y de la Vida, Universidad de Guadalajara, Centro Universitario de los Lagos, Enrique Diaz de León 1144, Col. Paseo de la Montaña, Lagos de Moreno C.P. 47460, Jalisco, Mexico; vmaranon@culagos.udg.mx
* Correspondence: scalixto@cio.mx; Tel.: +52-477-441-4200

Received: 4 December 2018; Accepted: 21 December 2018; Published: 21 January 2019

Abstract: The use of electronic devices to measure Relative Humidity (RH) is widespread. However, under certain circumstances, for example when explosive gases are present, a spark-free method should be used. Here we suggest the use of stimuli-responsive materials, like gelatin and interpenetrated polymers, to detect RH with an optical method. These materials are hydrophilic. When water vapor is absorbed by the films the molecules attach to the films molecular network. The result is that the film thickness increases and their refractive index changes. To detect the change of these two parameters an optical method based on diffraction gratings is employed. Surface diffraction gratings are recorded on the films. Then gratings are placed in an optical configuration that is immersed in a climatic chamber. A light beam is sent to the grating where it is diffracted. Several light orders appear. Due to the absorption of water molecules the films swell and grating surface modulation changes. This implies that the diffracted orders intensity changes. A calibrating plot relating intensity as a function of RH is obtained.

Keywords: stimuli-responsive materials; gelatin; interpenetrated polymers; relative humidity; diffraction gratings; permeability; climatic chamber

1. Introduction

Stimuli-responsive polymers present changes when subjected to external parameters like pH, temperature, solvent composition, radiation, mechanical stress, light, electrical and magnetic fields, and chemical triggers like glucose to mention but a few. These stimuli can trigger discontinuous changes (phase transitions) or gradual changes that occur over a finite range of stimulus levels. The changes could be reversible but hysteresis may occur. In recent years, new, smart polymers with dynamic properties were designed, by exploiting either reversible covalent bonds or supramolecular interactions to induce topological changes in chain structure [1,2].

In general stimuli-responsive materials can be used in sensors, drug delivery carriers, artificial muscles and more [3]. Among the key properties of stimuli-responsive materials for most applications is their degree of swelling, swelling kinetics and permeability. The swelling degree (i.e., the ratio of wet mass to dry mass) is important because it affects the kinetics, permeability and modulus. Among other factors, the swelling of stimuli-responsive materials is a function of crosslinking junction concentrations. Swelling kinetics and permeability are related. Increasing the swelling degree dilutes the network and this leads to an increased permeability to solutes. Conversely, a decrease of the swelling degree gives a decrease in permeability. Analysis of the swelling or shrinking kinetics of materials is primary for the design of the stimuli-responsive applications. Transparency can be a key for some applications, such as light sensors. Other key properties could be biocompatibility and biodegradability when stimuli-responsive materials have medical and pharmaceuticals applications.

The word humidity [4] is used when we are dealing with water vapor which is a gas. Water vapor is present in the earth's environment. Humidity measurements are important in industry, laboratory and common life activities. Thus humidity instrumentation that met rigid requirements has been developed. There are many types of humidity sensors called hygrometers. For every measurement the most useful sensor should be selected considering its high sensitivity, short response time, linear response, long term stability, small hysteresis, good durability, resistance against contaminants, cost, maintenance, calibration requirements, ease of installation and service to mention but a few. Besides this there are several techniques to measure humidity. Because there is no humidity sensor available that can cover the full dynamic range of water vapor levels different measurement methods and sensors have been developed each having certain advantages and limitations and suitable for some, but not all, applications. The most fundamental measurement technique of humidity is the chilled mirror hygrometer [5]. Other techniques comprise optical fibers [5], Micro Electromechanical Systems (MEMS) [6], hygrometers based on paper [7] and poly(vinylalcohol) (PVA), nanocomposites [8] containing graphene, among many others. Hygrometers have sensors that interact with water vapor by absorption, adsorption, and desorption of water molecules. Absorption could be by diffusion and capillary action.

In this paper we present a continuation of the work that we have developed and presented before [9,10]. In Section 2 we describe the materials used to develop the sensor in the form of films and the method to fabricate them. Section 3 shows the Relative Humidity (RH) characterization of films behavior. Section 4 describes the gratings theoretical background. Section 5 shows the gratings optical fabrication method and the study of gratings profile. Section 6 describes the climatic chamber and calibration method to quantify RH by means of the diffracted orders intensity. Section 7 deals with hysteresis shown by the films. Finally in Section 8 conclusions are drawn.

2. Materials

2.1. Gelatin, PVA and Poly(Acrylic Acid) (PAA)

Gelatin is a derived protein that does not occur free in nature. It is primarily used in the food and pharmaceutical industries as a gelling agent. The colorless gels that gelatin forms, in the presence of an appropriate solvent, are thermoreversible when cooled below about 35 °C.

For gelatin production, collagen from porkskins, cattle bones or hides is hydrolyzed and extracted either with an acid or alkaline pre-treatment, generating the Type A and Type B products, respectively. Both types of gelatin consist of a heterogeneous mixture of single or multi-stranded polypeptides containing less than 1000 amino acids. This is due to the partial breakage of collagen and the renaturation of the collagen-like helical structure among some of these fragments.

The aminoacidic chains of gelatin are typically enriched in glycine (approximately 30% of the aminoacidic content), proline and 4-hydroxyproline residues. These aminoacidic chains display several groups that can form hydrogen bonds with water, predominantly the hydroxyproline hydroxyl groups and the peptide carbonyl groups.

The properties of gelatin depend heavily on its water content: If dehydrated below a 2% moisture level, it becomes insoluble in water because of crosslinking between the gelatin macromolecules. Dehydrated gelatin is thus a crosslinked polymer, extremely brittle in the solid state. Gelatin undergoes transformation to a rubbery state at around 25% moisture content and at room temperature.

These and other physical properties of gelatin might be correlated to the different states in which water is present inside the gel. Similarly to what is found in other protein mixtures or polymers, water inside gelatin can be free, weakly bound (also called intermediate water) or tightly bound [11]. The degree of water molecules association with the chains of the polymer, in this case with collagen helices, determines also the physical behavior of water itself: The water molecules that establish the strongest bond with the polymer lay in close proximity to the aminoacidic chains and are unable to freeze, while intermediate water molecules interact less tightly with the chains and undergoes cold crystallization, i.e., freezes at temperatures below 0 °C. Amidst the aminoacidic chains, free water is able to move and freezes at 0 °C; it is the first to leave gelatin when dehydration occurs. When the water content drops below 15–20%, virtually no free water is present inside the polymer [12].

Chemical structure of gelatin films can be modified by means of ultraviolet light, heat and some chemicals. For example if a blend of gelatin and dichromates is illuminated with UV light, crosslinking will occur in the gelatin network. Regarding chemicals the use of formaldehyde can harden the gelatin. Commercial photographic fixing baths contain some chemicals like sodium or ammonium thiosulfate, acetic acid, sodium sulfite and potassium alum. This last chemical will form more crosslinks in the gelatin network preventing excessive gelatin swelling or softening.

Poly(vinyl alcohol) PVA is an atactic material that exhibits great crystallinity; in structural terms it has 1,3-diol bonds $(-CH_2-(OH)-CHCH_2-(OH)CH-)_n$. It has been widely used for its emulsifying and adhesive properties, but its main feature related to this work is its ability to absorb water, which is why it is considered a hydrogel. The water molecules act as plasticizers, which reduces the tensile strength of the polymer and increases its elongation allowing its resistance to tearing. The permeation of oxygen through the hydrogel is substantially facilitated when there is equilibrium of water content with PVA. This polymer absorbs and retains large amounts of water molecules [13].

Poly(acrylic acid) (PAA) was used as a crosslinking reagent because it has a functional carboxyl group $(-CH_2-(CO_2H)CH-)_n$ in each monomer unit to react with PVA. Some characteristics shared by both polymers include a high solubility in water and a high miscibility of PAA with PVA. Both polymers are widely commercialized and readily available. They can form a strong crosslinking by an ester bond between the hydroxyl group of PVA and the carboxyl group of PAA. The crosslinking is caused by strong hydrogen bonds between carboxylic acid of PVA and the hydroxyl groups of PAA. Finally, both polymers are good hydrogels agents [14].

2.2. Interpenetrated Polymers

Some materials are fragile and break easily. To strengthen them they are polymerized and crosslinked with other polymers [15,16]. Combination of the two polymers can be prepared in the form of blends, copolymers and interpenetrating polymer networks (IPNs) or interpenetrated hydrogels (IPHs). The IPNs are a combination of at least two polymer chains each in the network form. They form polymer gels held together by permanent entanglements. One polymer is synthesized or crosslinked independently of the presence of the other. Polymers are concatenated and cannot be pulled apart, but they are not bonded to each other by any chemical bond.

To fabricate the hydrogels the following chemicals, purchased from Sigma Aldrich (St. Louis, MO, USA), were used. The acrylic acid (AA) had a M_w 72 g/mol with a purity of 99%. The poly(vinyl alcohol) (PVA) had an M_w between 89,000–98,000 g/mol, 99% hydrolyzed. N-N' methylene bis (acrylamide) with an M_w 154 g/mole, purity 99%, was used as a crosslinker, and potassium persulfate was used as a thermal initiator. PVA/PAA IPN films were fabricated following the method described in references [17,18].

2.3. Thin Films Fabrication Method

The materials to be used as sensors should be available in transparent and thin film form. To make the films the surface of two glass plates of about 8.5 cm × 8.5 cm were polished and a hole of about 6 cm diameter was made in one plate, Figure 1a). The plates were placed in contact and the set was leveled. In the hole a mixture of gelatin and water (200 mg of gelatin in 10 mL of water) or PVA/PAA was poured, Figure 1b). After about 24 h a thin film was present. The amount of poured mixture is related to the film thickness. After the film was made a metallic ring was glued to the film, Figure 1c). Then the ensemble was detached from the glass.

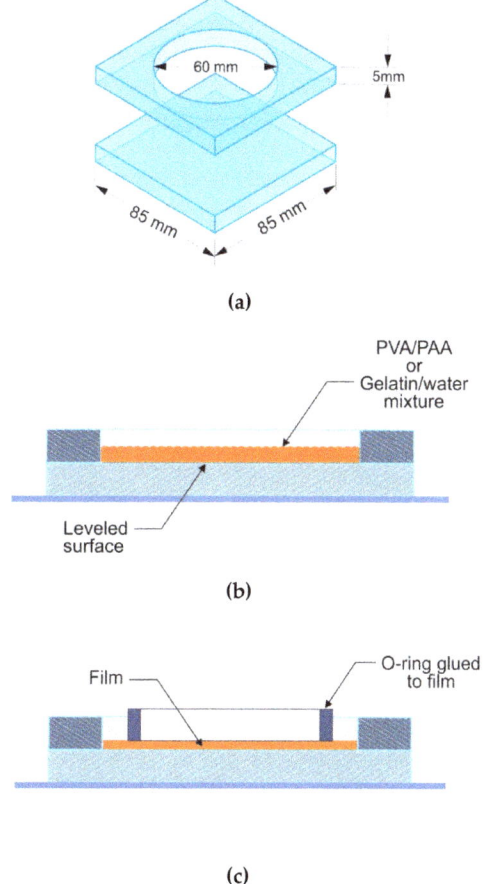

Figure 1. Schematic shows the process to make thin films: (**a**) two glass pieces; (**b**) pouring the mixture and (**c**) gluing the metallic ring.

3. Thin Films Characterization Methods Related with Water Absorption and Permeability

All the experiments described in this article were performed at room temperature. Films used as the active sensor for RH measurements were characterized for their water absorption and permeability characteristics.

3.1. Behavior of Films Weight as a Function of Water Molecules Absorption

Water absorption or desorption affects films weight. As we have seen in Section 2 when water molecules are absorbed by the films they interact with the chains of molecules that compose the films. The result is that the weight of the film increases. The opposite happens when the molecules are desorbed. Several materials were tested with the help of a microbalance.

Gelatin and PVA/PAA films were made with a diameter of 6 cm and different thicknesses. Then they were kept in a box that had silica gel for about two hours. After this they were placed in a microbalance and the behavior of their weight as a function of time was found. In the course of time water molecules in the atmosphere were trapped by the films and their weight increased.

In Figure 2 the behavior of the film's normalized weight as a function of time is shown, for two gelatin films with thickness of 50 μm and 15 μm and for a PVA/PAA film, with a thickness of 50 μm.

The gelatin film with 15 μm thickness absorbs quickly the water molecules and reaches the equilibrium state in about 300 s. The 50 μm gelatin film tends to stabilize in about 700 s and the PVA/PAA in about 2500 s.

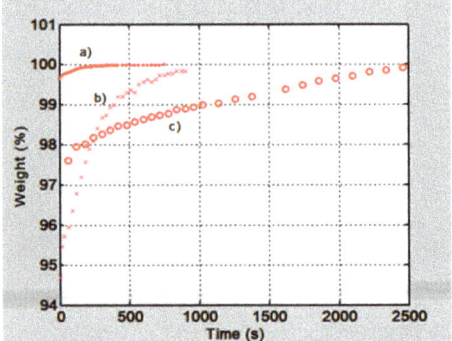

Figure 2. Normalized films weight as a function of time when they were placed in a microbalance and let them absorb environmental water molecules. Plot a) is for a gelatin film with a thickness of 15 μm, plot b) for a 50 μm film and plot c) for a poly(vinyl alcohol) (PVA)/poly(acrylic acid) (PAA), 50 μm film.

To find the behavior of gelatin films water absorption as a function of the film's thickness some films were made. They had the following thicknesses: 10, 20, 30, 40 and 50 μm. After fabrication they were kept in a box with silica gel for about two hours and then they were placed in the microbalance one at a time. During this weighing process they absorbed water molecules from the atmosphere. Plots like the ones shown in Figure 2a),b) were obtained. Then to know the amount of water that was absorbed by the films the weight values at the beginning and at the end of the process were taken from each plot. The one at the end was taken when the plot reached the equilibrium state. The difference between these two measurements gave us the amount of absorbed water, in mg. These values are plotted as a function of the film's thickness in Figure 3. We can notice that as the film thickness increases the film absorbs more water.

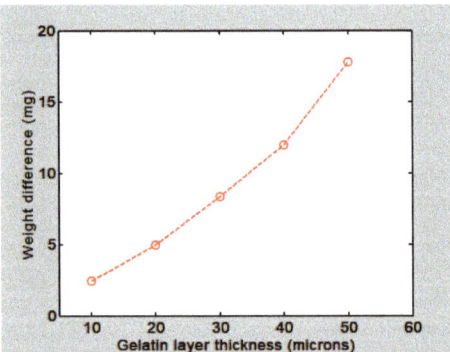

Figure 3. Plot shows the weight difference (mg) between the values when the films were dry and after they had absorbed water molecules from the atmosphere as a function of the film's thickness (μm).

3.2. Hardened and Unhardened Gelatin Films

It has been mentioned in Section 2.1 that gelatin thin films are hardened by means of UV and visible light or by means of chemical methods. Hardened films will be less hydrophilic because molecular chains will present more crosslinks. We have tested this statement in the following way.

A 15 µm thickness gelatin film was made and its water absorption behavior was tested by weighting the film through time. This behavior can be seen in Figure 4, plot a). Then the film was hardened with a photographic film hardener for 3 min [19]. A wash bath for 4 min followed. After drying, the film was placed in a container with silica gel for 2 h and then the weight test was done. The result is shown in Figure 4, plot b). We can notice that plot a), for unhardened film, reaches stability after about 450 s, however, plot b), for the hardened layer, reaches stability after 1200 s. Thus, the film affinity to water molecules has been reduced by hardening it or its sensitivity to RH was diminished.

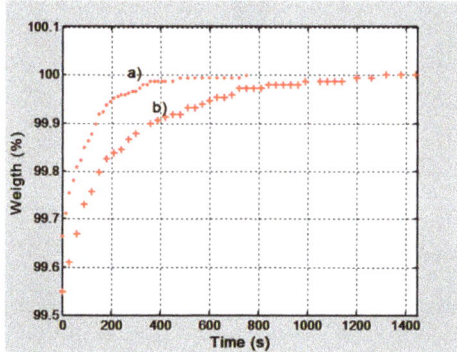

Figure 4. Behavior of the normalized weight of a gelatin thin film, 15 µm thickness, when it was in its unhardened state a) and hardened one b).

3.3. Films Water Vapor Permeability

In material science films permeability is tested through strict rules or methods. Some of them are mentioned in the ASTM E96 document (Standard tests methods for water vapor transmission of materials) [20]. This test method covers the determination of water vapor transmission (WVT) of materials through which the passage of water vapors may be of importance. Other methods have been exposed [21,22]. An alternative method to measure the films water vapor transmission was developed in our laboratory. A schematic of the device can be seen in Figure 5. An ordinary RH electronic gauge was coupled to a funnel. On top of the funnel a metallic ring was glued to the funnel with silicon. The metallic ring had a thin film glued to it. The funnel had an entrance hole and an exit one. Through the entrance dry air was pumped until a certain RH value (~11%) was reached. During this time the exit hole was opened, i.e., dry air passed freely taking out the water molecules. When the desired low RH was reached, pumping stopped and then the exit hole was closed.

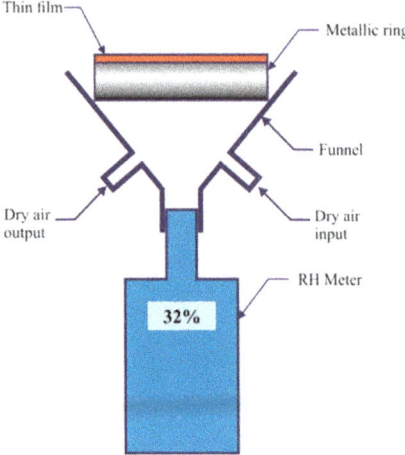

Figure 5. Schematic of the device used to find the film's water vapor permeability.

Through time the water vapor, present in the atmosphere, began to permeate through the film and the RH measurements were made. Plots in Figure 6 show the behavior of the RH as a function of time for two gelatin films. a) for a 15 µm thick and b) for 50 µm thick film. Moreover, a PVA/PAA film, with a thickness of 50 µm was tested, plot c). From the plots we can see that gelatin thin films let the water molecules pass more easily, while PVA/PAA films hinder the permeation of water molecules.

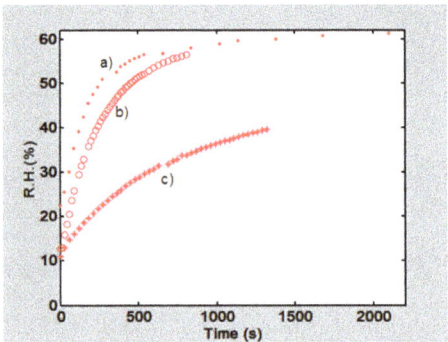

Figure 6. Behavior of Relative Humidity (RH) as a function of time when two gelatin films, a) and b), and a PAA/PVA film, c), were tested. Gelatin films presented thickness of 15 µm a) and 50 µm b). c) shows the behavior for a PVA/PAA, 50 µm film.

3.4. UV-Vis and FTIR-ATR Characterization

The UV-Vis and FTIR-ATR studies are applicable to the chemical characterization of polymers. It is a method that determines the crosslinking between polymers due to the formation of new bonds. The modes of vibration of the bonds between the atoms are differentiated according to the plots displacement in the wavenumber axis. The chemical structure of Gelatin, PAA, PVA and PVA/PAA films was determined with UV-Vis and FTIR spectra presented in Figures 7–10. The measurements were performed with a Lambda 365 and a Frontier spectrometer (Perkin Elmer, Waltham, MA, USA).

The UV-Vis spectra of the PVA/PAA films, with 30, 40 and 45 µm thickness, are shown in Figure 7. It is seen that when the thickness increases also the absorbance does. There is a maximum at 278 nm corresponding to the electronic transitions $n \rightarrow \sigma^*$, while the electronic transitions $\sigma \rightarrow \sigma^*$ are presented between 250–270 nm. The absorbance is a function of the films water molecules thickness inside the PVA/PAA films.

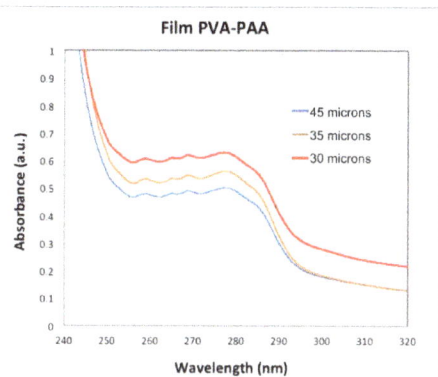

Figure 7. Absorbance as a function of wavelength for three PVA/PAA films with different thickness.

Table 1 presents the vibrational modes of bands given by gelatin, PVA, AA, and PVA/PAA films.

Table 1. Band assignment of IR vibrational modes for gelatin, PVA, acrylic acid (AA), and PVA/PAA films.

Entry	N–H st.	–O–H st.	H–O–H st.	C–H st.	=C–H st.	C=O st.	C=C st.	CH$_2$ b.	C–O b.	C–O st.	C–C st.
Gelatin	3793	3250	3595	2987	3125	1639	1535	1499	1480	1247	1090 1085
PVA	-	3347	3443	3000		-	-	-	1554	1244	1035 975
AA	-	3090	-	3000	2968 2902	1700	1434	-	1413	1298	1047 976
PVA/PAA 35 μm	-	3324	3443	3005	3094	1657–1618	1550	1340	1454	1244	1085 1035 976
PVA/PAA 30 μm	-	3324	3443	3005	3094	1657–1618	1550	1340	1454	1244	1085 1035 976

st. stretching; b. bending.

The FTIR-ATR spectrum of gelatin, Figure 8, shows the characteristic peak at 3250 cm^{-1} given by the hydroxyl group (–OH). The bending at 3595 cm^{-1} given by the amine group OH is also observed. At 3125 cm^{-1} the characteristic vibrations of (–CH) groups are seen. The 2987 cm^{-1} peak corresponds to the symmetric vibrations of the protein (C–H). The stretching vibration from aromatic atoms can be seen between 2392 cm^{-1}–1950 cm^{-1}. The peaks at the wavenumbers 3250 cm^{-1}, 1639 cm^{-1} and 1535 cm^{-1} are attributed to the free water molecules or amides III, II and I respectively. The bend at 1650 cm^{-1} corresponds to the CO stretching. The bend at 1499 cm^{-1} can be attributed to weak NH bending and C–N stretching of amide II and the bend at 1247 cm^{-1} to weak KC–N stretching and NH bending of amide II. Finally, the bends at 1090 cm^{-1} and 1025 cm^{-1} represent the weak bending and C–N stretching of amide III.

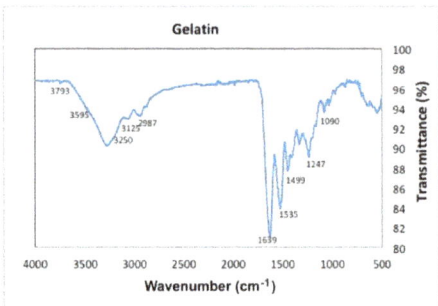

Figure 8. Transmittance as a function of wavenumber for gelatin.

Figures 9 and 10 show the FTIR-ATR spectra of PVA, PAA standards and the PVA/PAA films with a thickness of 30 μm and 35 μm. Figure 9 shows the wavenumber range between 3000 cm^{-1} and 2400 cm^{-1} and Figure 10 from 1800 cm^{-1} to 800 cm^{-1}. In Figure 9 the PVA plot shows a peak at 3347 cm^{-1} corresponding to the vibrations of hydroxyl bonds (–OH), which is not intense. For the AA plot at 3000 cm^{-1}, a wide and displaced signal appears and corresponds to the hydroxyl vibration modes (–OH) of the acid group. When the PVA/PAA crosslinked films are made, a very wide band centered at 3324 cm^{-1} appears and indicates the films ability to absorb water molecules into the polymer crosslinked network. As the thicknes of the PVA/PAA films increases, the range of water vibration modes (H–O–H) becomes wider and more intense (3443 cm^{-1}) due to the absorption of water molecules into the crosslinked network. In the 35 μm film the peaks at the wavenumbers 3000 cm^{-1}, 2968 cm^{-1}, and 2902 cm^{-1} are shown. These vibrations correspond to the asymmetric and symetric stretching vibrations of –CH. However, when the films are formed with the crosslinked PVA/PAA polymer, the vibrations are lost because the new C–C bonds of the interpenetrated hydrogel are formed at 2896 cm^{-1}. The plot for the 30 μm thick film shows the characteristic bond signals (=CH) of the vinyl group that appear at 3005 cm^{-1}. Finally it can be seen in the PVA/PAA plots that the thinner the film thickness the transmittance at around 3324 cm^{-1} decreases because there is a greater absorption of water molecules [23].

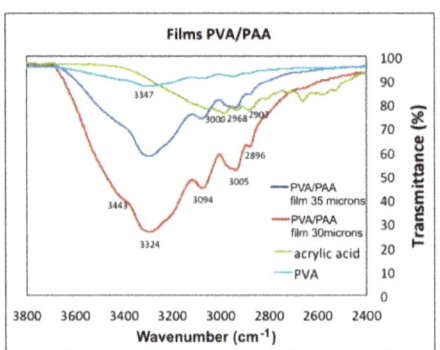

Figure 9. Transmittance as a function of Wavenumber for PVA, PAA, and two PVA/PAA films with different thickness.

Regarding the wavenumber range between 1800 cm^{-1} and 800 cm^{-1}, Figure 10, the PAA plot presents peaks at 1700 cm^{-1}, 1647 cm^{-1} and 1618 cm^{-1} that correspond to the vibration modes of carbonyls (C=O). The vibrations modes at 1298 cm^{-1} and 1244 cm^{-1} corresponds to the C–O bond. Concerning the PVA/PAA films the hydrogel intensifies the signal from 1657–1618 cm^{-1}, relative to the PAA plot, and a new signal appears at 1550 cm^{-1}, both signals correspond to the vibration modes of the new carbonyls (C=O) derived from the PAA forming the networks of cross-linking of the polymer as ester vibration. The peaks at 1657 and 1550 cm^{-1} are broad and intense because the water molecules are linked by blending the cross-linked PAA chain [24].

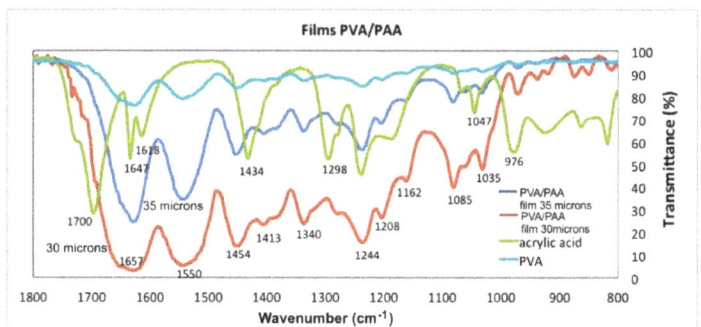

Figure 10. Transmittance as a function of wavenumber for PVA, PAA, and two PVA/PAA films.

4. Optical Principle of RH Sensing Method Using Stimuli-Responsive Polymer Systems. The Diffraction Grating

Theoretical grating behavior can be described by applying the thin phase screen approximation [25]. A grating with a sinusoidal profile is considered, Figure 11.

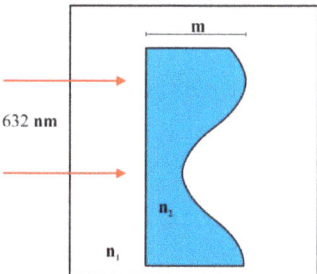

Figure 11. Sinusoidal grating profile.

The grating has a refractive index n_2 and it is immersed in a medium with refractive index n_1. Grating shows a deep modulation m. Illuminating light has a wavelength of 632 nm. After light passes the grating some spots appear due to the diffraction (diffracted orders). The first order intensity is described by the following formula:

$$I_1 \propto \left[J_1 \left(\frac{2\pi m}{2 \lambda 0} (n_1 - n_2) \right) \right]^2$$

In the present study the grating is immersed in the air, so $n_1 = 1$. PVA/PAA interpenetrating polymer and gelatin have refractive indices of 1.486 and 1.5, respectively. They were measured with an Abbe refractometer (Abbe refractometer, Bausch and Lomb, Rochester, NY, USA).

It has been mentioned that dry films absorb water molecules when RH increases. The result of this phenomenon is the film swelling and the decrease of its refractive index (n_2). Swelling causes the grating modulation (m) to increase. Considering the parameters m and n_2, the first order intensity behavior can be calculated by using the formula reported above. Results can be seen in Figure 12. The modulation (m) deep was varied from 0 to 3 µm and the parameter is the refractive index (n_2) that changed from 1.3 to 1.5. It is seen that when m = 0 the first order intensity is null because there is not a surface modulation. When modulation increases, first order intensity increases for all the plots, and presents a maximum intensity of 33%. Then it decreases and reaches a minimum value (zero). When the plot for $n_2 = 1.3$ is chosen its full width at half maximum (FWHM) is about $\Delta m = 1.28$. By taking two points in the linear part of the plot we can find the slope to be -0.3686. Following the same method, we can find the slope of the linear part of the plot when $n_2 = 1.5$ to be -0.59, and its FWHM $\Delta m = 0.769$. The slope of the plots shows us the change of intensity as a function of the change in modulation that is caused by changes in RH. This slope represents the sensitivity of the method. It is found that if refractive indices are closer to $n_2 = 1.5$ the sensitivity is greater than when refractive indices are close to 1.3 as seen in Figure 12.

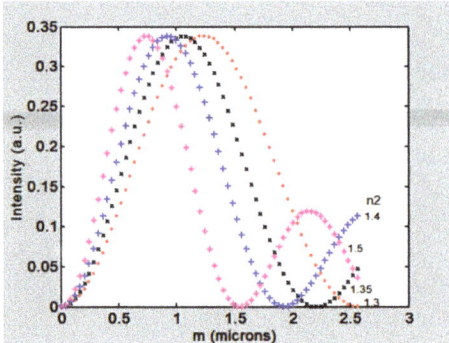

Figure 12. Theoretical behavior of the first order intensity as a function of grating relief modulation m. The refractive index n_2 presented several values.

5. Diffraction Grating Fabrication Method and Investigation of Grating Profiles

In previous work we used visible light to make the gratings [7]. Now a thermal method has been used to make the relief gratings [26]. Materials like gelatin and interpenetrated PVA/PAA films have been used. Thicknesses ranging from about 15 to 60 µm have been considered. To make the gratings a CO_2 laser ($\lambda = 10.6$ µm) has been used. A two beam interference configuration was set up, Figure 13. This resulted in an IR sinusoidal interference pattern. When the IR light was absorbed by the films the spots in the sinusoidal pattern, corresponding to high intensities/constructive interference, partially melted the film so that some material was redistributed into the adjacent spots of lower intensity. The result was a sinusoidal relief.

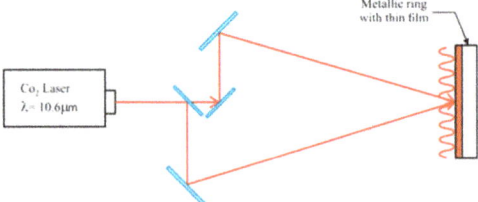

Figure 13. Infrared interferometric optical configuration to record the gratings on the surface of thin films. The sinusoidal red line represents the IR intensity field.

The gratings fabrication method considered parameters like thin films thickness, beam intensity (10 W), and exposure times. Photographs in Figure 14 show two gratings made with gelatin with two exposure times, 90 ms and 140 ms. It is possible to notice the fringes widening as the exposure time increases.

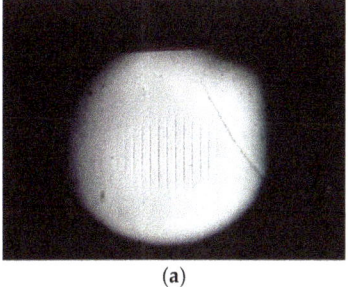 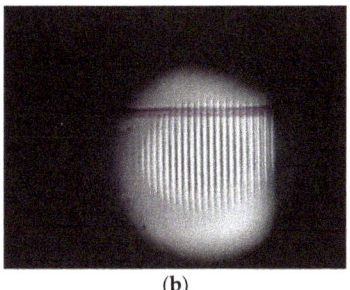

(a) (b)

Figure 14. Photographs showing two fabricated gratings with exposure times of (a) 90 ms and (b) 140 ms. The distance between fringes is 263 µm.

To investigate the gratings relief two instruments were used, a profilometer and an interference microscope. The first is a mechanical device and the second relies on light interference.

It has been mentioned that the consequence of the absorption of water molecules by the films, with any of the materials mentioned in Section 2, is their swelling. To find the effect of swelling on the grating profile we performed the following experiment. At first the grating was scanned with the tip of the profilometer. At this time the film was dry. Then the film was let absorb water molecules from the atmosphere. A second scanning was done. The result can be seen in Figure 15. Two plots are displayed. One shows the profile (b) when the film was dry and the other (a) when the film swelled. This last one shows a deeper relief.

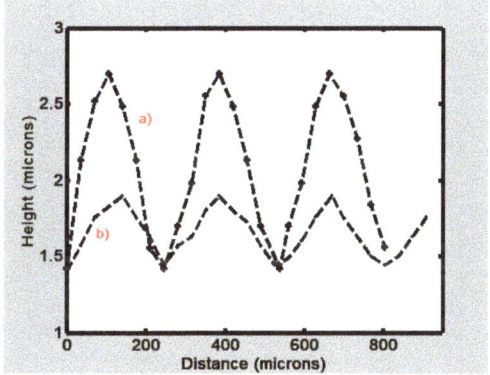

Figure 15. Profile height (µm) of a gelatin grating as a function of distance (µm). a) shows the profile after the gelatin thin film has absorbed water molecules. b) profile of a dry film.

Besides the investigation of the grating surface with the profilometer an interference microscope was used. This microscope is based on a Michelson Interferometer. In one light trajectory a plane mirror is placed. In the other trajectory the object to be investigated is inserted, the grating. Then at the output of the interferometer the two beams interfere. The interference pattern shows the relief of the object. In Figure 16 an image is seen when a grating made with PVA/PAA film was investigated. The distance between crests was 630 µm. Each interference line shows a sinusoidal profile.

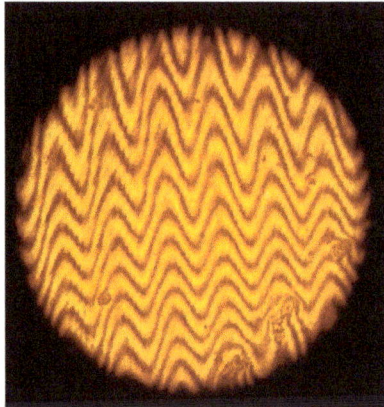

Figure 16. Gratings relief was investigated with an interference microscope. Photograph shows the relief of a PVA/PAA grating.

6. Humidity Detection with Diffraction Gratings

It has been seen in Section 4 the theoretical description of relief diffraction gratings and in Section 5 their fabrication methods. Here we describe the use of diffraction gratings to detect RH.

To test the films as RH sensors a means to control the RH is needed. This instrument is named a climatic chamber. Figure 17 shows a schematic of the simple chamber made in our laboratory. A transparent plastic box with a lid was connected to an air pump that removed the humid air from the box. Then a hollow plastic cylinder filled with silica gel absorbed the water molecules and the dry air leaving the cylinder was pumped back into the box. Two petri dishes were placed inside the box. One contained water and the other silica gel. Both dishes had lids that can be displaced from outside the box. The former dish was used to increase the relative humidity and the latter helped to decrease the humidity together with the air pump. An electronic RH sensor [27] inside the box measured the humidity in the chamber as a means of calibration. Some mechanical mounts inside the box allowed holding the metallic rings with the thin films glued to them.

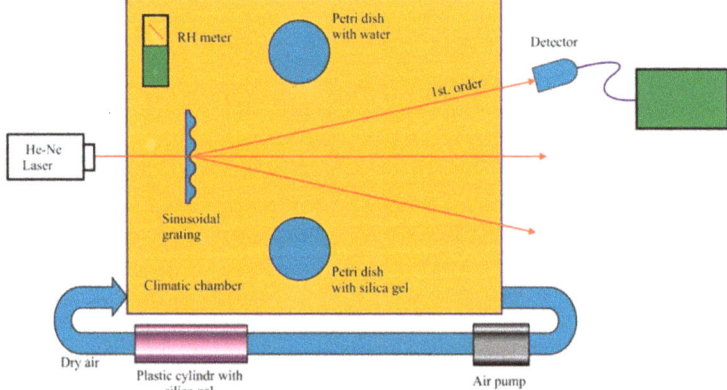

Figure 17. Schematic of a climatic chamber. Electronic RH instrument was used for calibration purposes. One Petri dish contained water and the other silica gel. Both dishes had lids that can be moved from outside.

An experiment was done to find the behavior of the RH when the petri dish lid, containing water, was displaced, Figure 18. It is seen that RH increases with time. At the end it tends to stabilize.

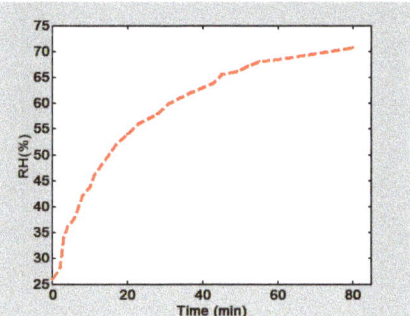

Figure 18. Behavior of RH in the climatic chamber as a function of time. RH was measured inside the climatic chamber when the petri dish lid that contained water was displaced.

The gratings were placed in the climatic chamber and a beam of light coming from an intensity stabilized He-Ne laser (λ = 632.8 nm) illuminated the grating. At the far field, outside the box, some light spots (diffracted orders) were seen, Figure 19. Dry air was pumped into the climate chamber until a desired low RH was reached. Then the lid of the petri dish, that con tained water, was opened. Humidity began to increase and the first order intensity changed due to the materials swelling. The plot in Figure 20 shows the behavior of the first order intensity as a function of the RH for a PVA/PAA film. It is possible to see that the response is not immediate. RH needs to increase to let the film swell. The slope of this plot is 0.08/RH. The RH range is from about 45% to 65%.

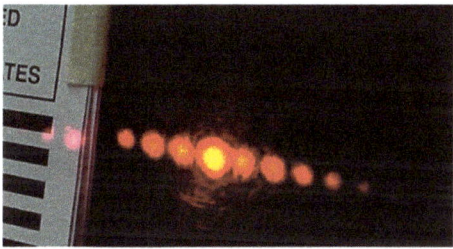

Figure 19. Diffracted orders given by a PVA/PAA grating.

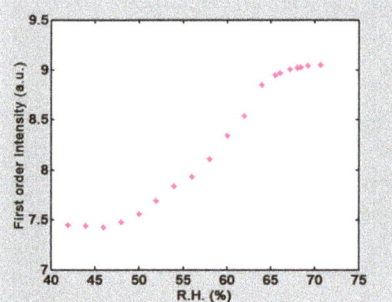

Figure 20. First order intensity as a function of RH in the climatic chamber. A PVA/PAA film with 50 μm thickness was used.

Besides the PVA/PAA film two gelatin films were tested. One had 15 µm thickness and the other 50 µm. Their normalized orders intensity behavior can be seen in Figure 21. For the 15 µm thickness film we have a slope of 0.91/RH, ranging from about 30% to 36% RH, and for the 50 µm thickness film a slope of 0.058/RH ranging from about 33% to 58% RH. Thus thin films are more sensitive than thick films but are useful in a shorter range. Thick films are less sensitive but useful in a wider range.

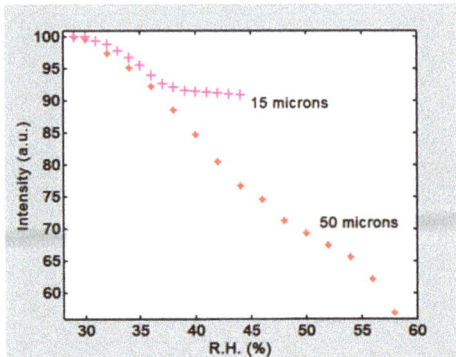

Figure 21. Normalized First order intensity, as a function of RH, for two gratings recorded in 15 µm and 50 µm thickness gelatin layers.

We should mention that when RH reached a value of about more than 70% the films became loose, Figure 22. At this stage the orders light intensity values should not be considered because it is possible that light illuminates another part of the grating that could have another profile.

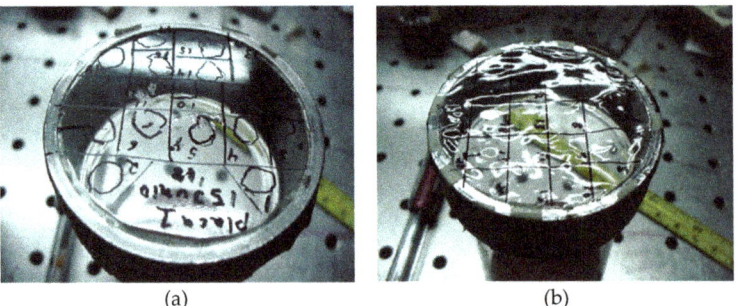

Figure 22. (**a**) shows the tense film surface when low humidity was present; (**b**) shows the film surface when humidity higher than about 70% was present. A loose surface can be seen.

7. Gels Hysteresis

The response of gelatin thin films when RH increased and decreased was investigated. A gelatin thin film with 15 µm thickness was used. First RH was increased by opening the petri dish that contained water in the climatic chamber. Then at a given RH it was closed and the petri dish that contained silica gel was opened. The behavior of the first order intensity can be seen in Figure 23.

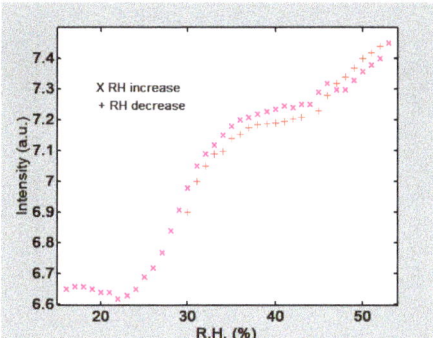

Figure 23. Behavior of first order intensity as a function of RH when it increases and decreases. Thin film thickness 15 µm.

8. Conclusions

It has been presented the feasibility of using stimuli-responsive materials, like gelatin and PVA/PAA, in optical hygrometers. A study has been presented before [28] where they used gelatin as RH sensitive material. However, they considered the use of electronic devices to measure the capacitance of the gelatin films and not the optical and mechanical characteristics of films, refractive index and swelling, like we have done. Besides, the fabrication of those sensors [28] requires high tech instruments and methods like photolithography, spin coating, thermal physical evaporation and wet etching techniques. Characterization studies presented now comprise the behavior of gelatin and PVA/PAA films permeability and weight as a function of RH. Additionally diffraction gratings have been generated on the film's surface. These gratings were placed in an optical configuration where light illuminated them and diffracted orders appeared. When RH surrounding the gratings changed the diffracted orders intensity also did it. Thus we have a calibration plot relating the light intensity as a function of RH.

These results should be taken as preliminary ones. Additional studies should be done by interdisciplinary research teams of chemist, physicists and materials science scientists to improve the response of films when used in hygrometers. In depth gelatin studies [29,30] should be considered.

A method to improve the sensitivity of the films to RH could be done by incorporating macropores in the films to let water molecules to be absorbed quickly. Also the use of PVA/gelatin interpenetrated films can be considered [31].

Regarding the Limit of Detection (LOD) of the proposed method we have the following comment. In Section 6 it was exposed that the slopes for plots in Figures 20 and 21 (Intensity vs RH) were 0.08/RH and 0.091 /RH, If we suppose that the light detector can measure a minimum intensity with a value of 0.1 µW then the LOD, considering the slopes, will be 0.125 RHU. (Relative Humidity Unit) and 0.109 RHU respectively. However, it should be remembered that the method presented here can be modified according to the experimental conditions. We have variables that can be adapted to the needs. For example the RH sensitive materials, gelatin and interpenetrated polymers, can be modified. Besides the grating parameters can be changed. We have seen in Section 4 that it is possible to select the grating refractive index and this in turn will change the sensitivity of the method. Besides this it is possible to choose more powerful lasers and light detectors with greater sensitivities. Thus, given the RH measurement it is possible to select the material and the optical parameters that will better suit the experiment.

In the market there are many types of hygrometers. Each is useful for certain applications. For example we have mentioned one made with paper [7]. Also, there are others made with graphene oxide [8,32]. The first one relies on capacitance measurements and has a sensitivity of 2 pF/RH%. The one in reference [32] is based on frequency measurements and has a sensitivity of 719 Hz/RH%.

The paper sensor is relatively easy to make. The GO sensor needs expensive, dedicated instruments, and expensive materials, and the fabrication method is difficult. The comparison of the RH sensor that we suggest with other sensors is difficult to make because it depends mainly on the application.

By means of the optical method presented here it should be possible to study the response of materials to RH.

Author Contributions: Conceptualization, S.C.; Gelatin paragraph, V.P.; UV and FTIR-ATR films characterization, V.F.M.-R.; Experiments, S.C.; Writing—original draft preparation, S.C., V.P., V.F.M.-R.; Writing—review and editing, S.C., V.P., V.F.M.-R.; Funding acquisition, S.C., V.P.

Funding: This research received no external funding.

Acknowledgments: We would like to thank Maria Elena Calixto-Olalde and Sachenka Bravo-Gomez for the PVA/PAA films fabrication. Also we thank Guillermo Garnica for fruitful discussions and Raymundo Mendoza for some drawings.

Conflicts of Interest: The authors declare no conflict of interest.

References

1. Sun, H.; Kabb, P.C.; Dai, Y.; Hill, M.R.; Ghiviriga, I.; Bapat, A.P.; Sumerlin, B.S. Macromolecular metamorphosis via stimulus-induced transformations of polymer architecture. *Nat. Chem.* **2017**, *9*, 817–823. [CrossRef]
2. Sun, H.; Kabb, P.; Sims, M.B.; Sumerlin, B.S. Architecture-transformable polymers. Reshaping the future of stimuli responsive polymers. *Prog. Polym. Sci.* **2018**. [CrossRef]
3. Yui, N.; Mrsny, R.J.; Park, K. (Eds.) *Reflexive Polymers and Hydrogels*; CRC Press: Boca Raton, FL, USA, 2004.
4. Wiederhold, P. *Water Vapor Measurements: Methods and Instrumentation*; Marcel Dekker: New York, NY, USA, 1997.
5. Yeo, T.L.; Sun, T.; Grattan, K.T.V. Fibre-Optic sensor technologies for humidity and moisture mesaurements. *Sens. Actuators A* **2008**, *144*, 280–295. [CrossRef]
6. Fenner, R.; Zdankiewicz, E. Micromachined water vapor sensors: A review of sensing technologies. *IEEE Sens. J.* **2001**, *1*, 309–317. [CrossRef]
7. Gaspar, C.; Olkkonen, J.; Passoja, J.S.; Smolander, M. Paper as Active Layer in Inkjet-printed Capacitive Humidity Sensors. *Sensors* **2017**, *17*, 1464. [CrossRef] [PubMed]
8. Lim, M.-Y.; Shin, H.; Shin, D.M.; Lee, S.-S.; Lee, J.-C. Poly(vynil alcohol) nanocomposites containing reduced graphene oxide coated with tannic acid for humidity sensors. *Polymer* **2004**, *84*, 89–98. [CrossRef]
9. Calixto, S.; Andres, M.V. Water Vapor Sensors Based on the swelling of Relief Gelatin Gratings. *Adv. Mater. Sci. Eng.* **2015**, *2015*, 1–5. [CrossRef]
10. Calixto, S.; Calixto-Olalde, M.-E.; Hernandez-Barajas, J.; Vazquez-Espitia, O. Mach-Zehnder interferometer applied to the study of polymer's Relative Humidity Response. In Proceedings of the MOEMS and Miniaturized Systems, San Francisco, CA, USA, 27 January–1 February 2018.
11. Tanaka, M.; Hayashi, T.; Morita, S. The roles of water molecules at the biointerface of medical polymers. *Polym. J.* **2013**, *45*, 701–710. [CrossRef]
12. Otsuka, Y.; Shirakashi, R.; Hirakawa, K. Bound states of water in gelatin discriminated by near-infrared spectroscopy. *Jpn. J. Appl. Phys.* **2017**, *56*, 111602. [CrossRef]
13. Sahoo, P.K.; Rana, P.K.; Swain, S.K. Interpenetring Polymer Network PVA/PAA Hydrogels. *Int. J. Polym. Mater. Polym. Biomater.* **2006**, *55*, 65–78. [CrossRef]
14. Mansur, A.; Mansur, H.; González, J. Enzyme-Polymers Conjugated to Quantum-Dots for sensing Applications. *Sensors* **2011**, *11*, 9951–9972. [CrossRef] [PubMed]
15. Billmeyer, F.W. *Textbook of Polymer Science*; John Wiley and Sons: New York, NY, USA, 1984.
16. Odian, G. *Principles of Polymerization*; John Wiley and Sons: New York, NY, USA, 1991.
17. Hernandez, R.; Lopez, D.; Mijangos, C. Preparation and characterization of polyacrylic Acid-Poly (Vinyl alcohol)—Based interpenetrating hydrogels. *J. Appl. Polym. Sci.* **2006**, *102*, 5789–5794. [CrossRef]
18. Byun, J.; Lee, Y.-M.; Cho, C.-S. Swelling of thermosensitive interpenetrating polymer networks composed of poly(vinyl alcohol) and poly(acrylic acid). *J. Appl. Polym. Sci.* **1996**, *61*, 697–702. [CrossRef]
19. Kodak. Available online: https://www.kodak.com/US/en/corp/default.htm (accessed on 28 December 2018).

20. ASTM American Society for Testing and Materials. *ASTME96 Standard Test Methods for Water Vapor Transmission of Materials*; ASTM American Society for Testing and Materials: Conshohockon PA, USA, 2016.
21. Schmidt, H.; Marcinkowska, D.; Cieslak, M. Testing Water Vapour Permeability Through Porous Membranes. *Fibres Text. East. Eur.* **2005**, *13*, 66–68.
22. Chambi, H.N.M.; Grosso, C.R.F. Mechanical and water permeability properties of biodegradables films based on methylcellulose, glucomannan, pectibn and gelatin. *Food Sci. Technol.* **2011**, *31*, 739–746.
23. Arndt, K.F.; Ritcher, A.; Zimmermann, J.; Kressler, J.; Kickling, D.; Adler, H.J. Poly(vinyl alcohol)/poly(acrylic acid) hydrogels: FT-IR spectroscopic characterization of crosslinking reaction and work at transition point. *Acta Polym.* **1999**, *50*, 383–390. [CrossRef]
24. Li, B.; Lu, X.; Yuan, J.; Zhu, Y.; Li, L. Alkaline poly(vinyl alcohol)/poly(acrylic acid) polymer electrolyte membrane for Ni-MH battery application. *Ionics* **2014**, *21*, 141–148. [CrossRef]
25. Calixto, S.; Bruce, N.C.; Rosete-Aguilar, M. Diffraction grating-based sensing optofluidic device for measuring the refractive index of liquids. *Opt. Express* **2015**, *24*, 180–190. [CrossRef]
26. Calixto, S. Infrared recording with gelatin films. *Appl. Opt.* **1988**, *27*, 1977–1983. [CrossRef]
27. Extech Instruments Corporation. *Moisture Meter Model MO210*; Extech Instruments Corporation: Waltham, MA, USA.
28. Sharpardanis, T.; Hudpeth, M.; Kaya, T. Gelatin as a new humidity sensing material: Characterization and limitations. *AIP Adv.* **2014**, *4*, 127–132.
29. Kozlov, P.V.; Burdygina, G.I. The structure and properties of solid gelatin and the principles of their modification. *Polymer* **1983**, *24*, 651–656. [CrossRef]
30. Abd El-Kader, P.F.H.; Gafer, S.A.; Basha, A.F.; Bannan, S.I.; Basha, M.A.F. Thermal and Optical Properties of Gelatin/Polyvinylalcohol Blends. *J. Appl. Polym. Sci.* **2010**, *118*, 413–420. [CrossRef]
31. Hago, E.-E.; Lin, X. Interpenetrating Polymer Network Hydrogels: Synthesis and Characterization. *Adv. Mater. Sci. Eng.* **2103**, *2013*, 328763.
32. Sun, C.; Shi, Q.; Yasici, M.S.; Lee, C.; Liu, Y. Development of a highly sensitive humidity sensor based on a piezoelectric micromachined ultrasonic transducer array functionalized with grapheme oxide thin film. *Sensors* **2018**, *18*, 4352. [CrossRef] [PubMed]

© 2019 by the authors. Licensee MDPI, Basel, Switzerland. This article is an open access article distributed under the terms and conditions of the Creative Commons Attribution (CC BY) license (http://creativecommons.org/licenses/by/4.0/).

Article

4D Printing Self-Morphing Structures

Mahdi Bodaghi [1,*], Reza Noroozi [2], Ali Zolfagharian [3], Mohamad Fotouhi [4] and Saeed Norouzi [2]

[1] Department of Engineering, School of Science and Technology, Nottingham Trent University, Nottingham NG11 8NS, UK
[2] School of Mechanical Engineering, Faculty of Engineering, University of Tehran, Tehran, Iran; reza.noroozi@ut.ac.ir (R.N.); saeednorouzi@ut.ac.ir (S.N.)
[3] School of Engineering, Deakin University, Geelong, Victoria 3216, Australia; a.zolfagharian@deakin.edu.au
[4] Department of Design and Mathematics, the University of the West of England, Bristol BS16 1QY, UK; mohammad.fotouhi@uwe.ac.uk
* Correspondence: mahdi.bodaghi@ntu.ac.uk; Tel.: +44-115-84-83470

Received: 29 March 2019; Accepted: 22 April 2019; Published: 25 April 2019

Abstract: The main objective of this paper is to introduce complex structures with self-bending/morphing/rolling features fabricated by 4D printing technology, and replicate their thermo-mechanical behaviors using a simple computational tool. Fused deposition modeling (FDM) is implemented to fabricate adaptive composite structures with performance-driven functionality built directly into materials. Structural primitives with self-bending 1D-to-2D features are first developed by functionally graded 4D printing. They are then employed as actuation elements to design complex structures that show 2D-to-3D shape-shifting by self-bending/morphing. The effects of printing speed on the self-bending/morphing characteristics are investigated in detail. Thermo-mechanical behaviors of the 4D-printed structures are simulated by introducing a straightforward method into the commercial finite element (FE) software package of Abaqus that is much simpler than writing a user-defined material subroutine or an in-house FE code. The high accuracy of the proposed method is verified by a comparison study with experiments and numerical results obtained from an in-house FE solution. Finally, the developed digital tool is implemented to engineer several practical self-morphing/rolling structures.

Keywords: 4D printing; shape memory polymer; self-morphing; experiments; FEM

1. Introduction

In recent years, three-dimensional (3D) printing has dramatically been developed in various industrial fields to construct structures with complicated 3D shapes based on computer-aided design (CAD) models [1]. The process of creating 3D objects was invented in 1986 by Charles Hull and introduced as additive manufacturing (AM), rapid prototyping (RP), or solid-freeform (SFF) [2]. This convenient technology could construct 3D structures with thermoplastic polymer materials such as acrylonitrile butadiene styrene (ABS) [3–5], polylactic acid (PLA) [3,5,6], polyamide (PA) [7], and polycarbonate (PC) [8], that were already being used for biomechanics [9], optical metamaterials [10], smart textiles [11], and other applications. The advantages of this fabrication method are the optimal use of the material, a flexible design, and more precise production of complex parts and components.

As a class of multi-scale structures, so-called "metamaterials" exhibit thermo-mechanical properties that are not found in nature. Their unusual characteristics arise from their structures and geometries rather than the material of which they are composed [12]. For the first time, Lakes [13] reported foam structures with negative Poisson ratios. Recently, 3D printing technology has enabled us to fabricate cellular materials with complex architectures [14]. For example, Wang et al. [15] showed dual-material auxetic metamaterials consisting of two parts, stiff walls and elastic joints, that did not show any instability during deformation. The finite element (FE) and experimental results showed that these metamaterials had distinctly different auxeticities and mechanical properties from traditional single-material auxetic metamaterials. In another case, Garcia and et al. [8] designed an all-dielectric uniaxial anisotropic metamaterial and then fabricated and tested it. It was manufactured from polycarbonate using a fused deposition modeling (FDM) 3D printing. Mirzaali et al. [16] used computational models and advanced multi-material 3D printing techniques to rationally design and additively manufacture multi-material cellular solids for which the elastic modulus and Poisson's ratio could be independently tailored in different directions. Yang et al. [17] used the classical planar tessellation theory to find regular 2D figures that can be used as configurations for first- and second-order honeycombs, and systematically explored the configuration characteristics of the existing two-dimensional (2D) and 3D auxetic and non-auxetic structures. Then, based on a topology analysis, they designed and classified 3D hierarchical metamaterials according to first-order and second-order configurations, which have tailored different ranges of Poisson's ratio and Young's modulus. Bodaghi et al. [18] conducted experimental and numerical studies on the mechanical behaviors of metamaterials made of hyperelastics under both tension and compression in a large strain range. They used the FDM method for 3D printing samples and explored metamaterial behaviors in tension and compression modes, revealing buckling instability characteristics.

By printing "smart" materials, 3D printing shifts to another level that is called 4D printing. In other words, 4D printing is a combination of 3D printing with time as its fourth dimension [19]. Figure 1 shows the difference between 3D and 4D printings, where in a 4D-printed sample, stimuli like water, heat, a combination of heat and light, and a combination of water and heat trigger actuations. The selection of the stimulus depends on the requirements of the specific application, which also determine the types of smart materials employed in 4D-printed structures. Among active materials, shape memory polymers (SMPs) have more advantages. These advantages are higher recoverable strain of up to 400%, lower density, lower cost, simple procedure for programming of shapes, and good controllability over the recovery temperature. SMPs can hence be utilized in the automotive and aerospace industries, and other fields. For example, Tibbits et al. [20] created a linear strip structure composed of rigid and active materials. This structure could transform into a corrugated structure when it was placed in water. This was a demonstration of 1D-to-2D shape-shifting using the self-bending mechanism. They printed a 2D flat surface with different swelling activated in the water that could change into closed-surface cubes with a 3D shape. Zhang et al. [21] showed that the release of the internal strain in printed samples generated during the 3D printing process made the printed structure remain flat under heating, and when it was cooled to room temperature, it changed into a 3D structure. Jamal et al. [22] showed a configuration change for tissue engineering where a 2D planar bio-origami changed into a 3D pattern by the self-bending operation. This configuration change was enabled by different swelling ratios of hydrogels and rigid materials in the water. By understanding the FDM printing method and SMP cycle, Bodaghi et al. [23] manufactured adaptive metamaterials enabled by functionally graded (FG) 4D printing technology, without application of any programming process and external manipulation. They implemented an in-house FE code to solve constitutive governing equations of SMP structures. Using 4D printing, Joanne et al. [24] studied cross-folding origami structures that were made of multi-material components along different axes and different horizontal hinge thicknesses with a single homogeneous material. Chen et al. [25] demonstrated geometrically reconfigurable, functionally deployable, and mechanically tunable lightweight metamaterials utilizing 4D printing. They introduced metamaterials that were made of photo-crosslinkable and temperature-responsive SMPs. By implementing 4D printing technology, Bodaghi and Liao [26] introduced tunable continuous-stable metamaterials with reversible thermo-mechanical memory operations. Zolfagharian et al. [27] provided a control-oriented modelling approach for 3D-printed polyelectrolyte soft actuators. They developed an electro-chemo-mechanical model for the 3D-printed polyelectrolyte soft actuators and validated it with the experimental data. A new class of metamaterials, so-called "self-morphing structures", has recently been introduced and studied [28]. Yu et al. [29] designed a new concept of a morphing wing based on SMPs and their reinforced composites. Tao et al. [30] simulated self-folding SMP hinged shells by implementing a complicated user defined material (UMAT) subroutine into the commercial FE software package of Abaqus.

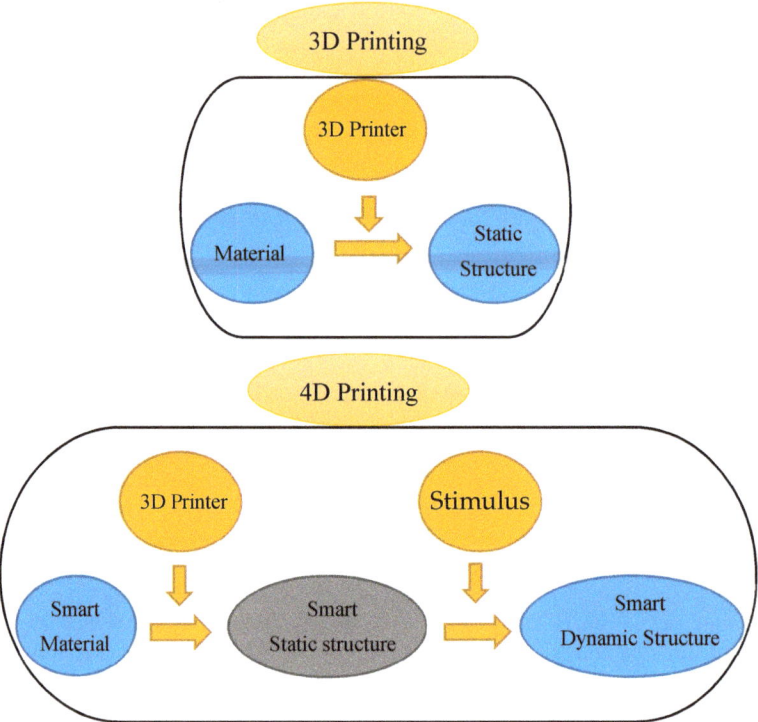

Figure 1. The difference between 3D and 4D printings.

The literature review implies that researchers have mostly developed their own in-house FE codes or implemented complicated UMATs using FE commercial software packages to model self-folding structures.

This paper aims at introducing self-bending/morphing/rolling structures fabricated by 4D printing technology and simulating their thermo-mechanical behaviors by a novel simple computational tool. The main approach is based on an understanding of thermo-mechanical behaviors of shape memory polymers and the concept behind FDM technology, as well as experiments to explore how printing speed can control self-bending features. The feasibility of the SMP primitives with self-bending features via FG 4D printing is first demonstrated experimentally. The self-folding 1D-to-2D process is simulated by introducing a straightforward method into a commercial FE software package of Abaqus that is much simpler than writing a UMAT subroutine or an in-house FE solution. 4D printing and the computational tool are applied to develop practical complex structures with self-bending/morphing/rolling features. A good qualitative and quantitative correlation is observed, verifying the accuracy of the proposed method.

2. Conceptual Design

2.1. FG 4D Printing

In this section, inspired by SMP features, we show the potentials of 3D printing in the design and development of the adaptive metamaterials. SMPs can retain a temporary shape and recover into their original shape when subjected to an environmental stimulus such as heat. Figure 2 shows an SMP thermo-mechanical cycle. The polymer is initially at temperature T_r. First, the polymer is heated up to temperature T_h that is greater than the glassy temperature of the polymer ($T_h > T_g$). Then, the material undergoes a strain level of ε_0 due to the applied load (loading). Next, it is fixed in the strain ε_0 while reducing the temperature to T_L, which is lower than T_g (cooling process). Afterwards, at the constant temperature T_L, the mechanical constraint is released (unloading). Upon heating, the pre-strain releases and the permanent shape is recovered (this is called stress-free strain recovery).

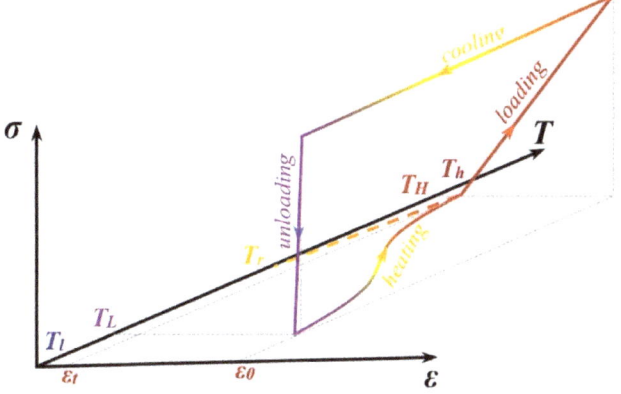

Figure 2. Shape memory polymer (SMP) thermo-mechanical cycle.

Among various 3D printing technologies, FDM technology applies a similar thermo-mechanical process to the material during printing. Figure 3 shows a schematic of the FDM method. The material is heated inside the liquefier at a temperature (T_{in}) that is greater than its T_g. It is then placed onto the platform by the printer head moving at the speed S_p. In fact, in this process, the material is stretched at a high temperature similar to the heating–loading process described for the SMP (heating and loading) producing a pre-strain. After placing each layer on the platform, the printed layer is cooled and solidified. This stage is like the SMP cooling step. Once a layer is printed, the build platform advances downward and the printing head proceeds to place the next layer. Finally, the thermo-mechanical programming process is completed by removing the 3D-printed object from the platform (mechanical unloading).

It is worth noting that, when printing the second layer, the hot material is placed on the first layer and reheats it. By partially reheating the first layer, some of the pre-strain is released. In a similar way, by printing other layers, the bottom layers are always reheated at each stage of printing and some of their pre-strains are released. Therefore, the first layer has the least pre-strain and the last layer has the maximum pre-strain as it never gets any heat, since the nozzle leaves it at the end of the 3D printing step. It can therefore be concluded that the printing speed can be considered as a control parameter that affects pre-strain values in the printed layers. This programming is performed in an FG manner, as the pre-stain can be changed layer-by-layer gradually.

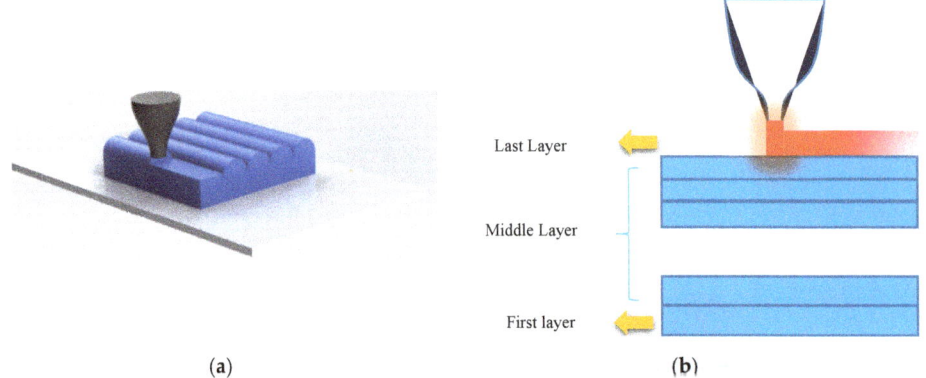

Figure 3. Schematic of the fused deposition modeling (FDM) method.

2.2. Materials and Printing

In this study, PLA filaments with diameters of 1.75 mm and glass transition temperatures of 63 °C are used. All the samples are manufactured with 3D printing and the FDM method. The extruder diameter is 0.4 mm and the liquefier temperature is set at 190 °C. The temperatures of the platform and chamber are set at room temperature and 24 °C, respectively. The 3D printing is performed at three different speeds, namely 20, 40, and 70 mm/s. The layer height is set to 0.1 mm.

The thermo-mechanical properties of the PLA are provided in this section. Dynamic mechanical analysis (DMA) (Q800 DMA, TA Instruments, New Castle, DE, USA) is performed to specify temperature-dependent material properties of the PLA. For this purpose, the sample is printed in the dimensions of 30, 1.6, and 1 mm for length, width, and thickness, respectively. Figure 4 shows the geometry of the 3D-printed sample and print direction.

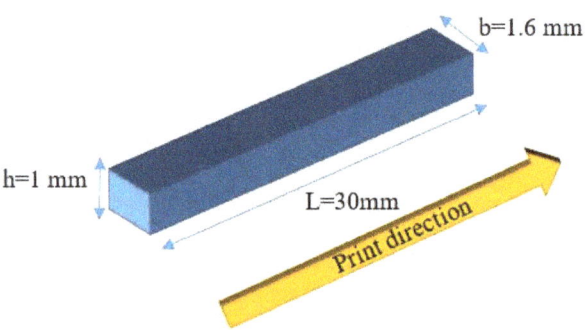

Figure 4. Geometry of the 3D-printed sample.

A DMA test is performed in an axial tensile condition with 1 H of force oscillation frequency and a 5 °C/min heating rate, with the temperature ranging from 30 to 93 °C. The results of the DMA test in terms of storage modulus E' and $\tan(\delta)$ are shown in Figure 5a,b, respectively. Also, the numerical results of the DMA test are presented in Table 1.

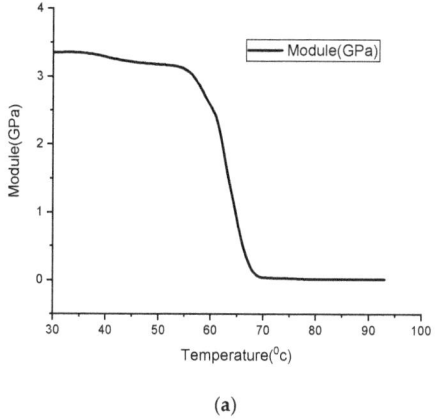

(a)

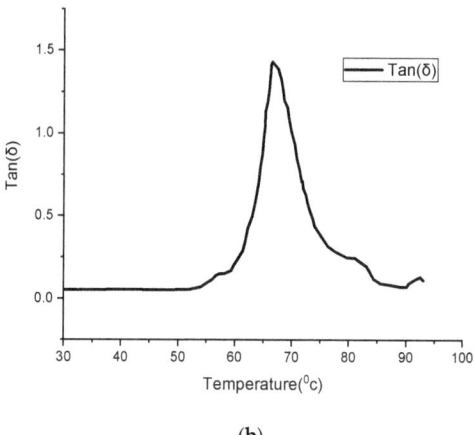

(b)

Figure 5. Dynamic mechanical analysis (DMA) results for polylactic acid (PLA): (**a**) Storage modulus E′, (**b**) tan(δ).

Table 1. DMA results for PLA.

T (°C)	30	40	50	52	54	56	60	62	64	66	68
Storage module	0.0524	0.055	0.0540	0.0561	0.0701	0.1134	0.2116	0.383	0.6753	1.2890	1.2295
tan(δ)	3.3510	3.288	3.1880	3.1682	3.1402	3.0777	2.6378	2.0433	1.2807	0.5393	0.1459
T (°C)	70	72	74	76	78	80	82	84	88	90	93
Storage module	0.9875	0.658	0.434	0.3292	0.2758	0.2483	0.221	0.1398	0.0774	0.0723	0.1248
tan(δ)	0.030	0.026	0.0218	0.0138	0.0117	0.0117	0.0117	0.0117	0.0117	0.0117	0.030

The peak in the graph of tan(δ) shows the glass transition temperature that is read as $T_g = 66$ °C.

3. Theoretical Modeling

3.1. SMP Model

Shape memory polymers are a new class of materials that can keep a temporary shape and return to their original shape upon application of a stimulus such as temperature. They consist of glassy and rubbery phases. Thus, we can show the volume fraction of the rubbery and glassy phases by scalar variables ξ_g and ξ_r as:

$$\xi_g = \frac{V_g}{V} \xi_r = \frac{V_r}{V} \tag{1}$$

where V_g is the glassy phase volume and V_r denotes the rubbery phase volume. The summation of the volume fractions of the two phases should equal to unity ($\xi_g + \xi_r = 1$). The transformation of the rubbery phase into the glassy phase is considered to be only a function of temperature. Thus, ξ_g and ξ_r only depend on the temperature. The volume fraction of the rubbery phase can be written in terms of the glassy volume fraction as:

$$\xi_r = 1 - \xi_g \tag{2}$$

We also assume that ξ_g is an independent variable and can be expressed as:

$$\xi_g = \xi_g(T) V_g = V_g(T) \tag{3}$$

Considering the experimental DMA test results, the glassy volume fraction can be interpolated by an explicit function as:

$$\xi_g = \frac{\tanh(a_1 T_g - a_2 T) - \tanh(a_1 T_g - a_2 T_h)}{\tanh(a_1 T_g - a_2 T_h) - \tanh(a_1 T_g - a_2 T_l)} \tag{4}$$

where a_1, a_2 are chosen to fit the DMA curve.

The rubbery and glassy phases in SMPs are assumed to be linked in series. Considering a small strain regime, justified by the fact that the printed structures experience small strains and moderately large rotations, additive strain decomposition is adopted as:

$$\varepsilon = \xi_g \varepsilon_g + (1 - \xi_g)\varepsilon_r + \varepsilon_{in} + \varepsilon_{th} \tag{5}$$

where ε denotes total strain; ε_g and ε_r indicate strain of the glassy and rubbery phases, respectively; ε_{in} is the inelastic strain due to phase transformation; and ε_{th} denotes the thermal strain which is defined as:

$$\varepsilon_{th} = \int_{T_0}^{T} \alpha_e(T) \, dT \tag{6}$$

where T_0 is the reference temperature and α_e is the equivalent thermal expansion, defined as:

$$\alpha_e = \alpha_r + (\alpha_g - \alpha_r)\xi_g(T) \tag{7}$$

During the cooling process, the rubbery phase transforms into the glassy phase and its strain, ε_{in}, is stored in the material. It is formulated as:

$$\dot{\varepsilon}_{in} = \dot{\xi}_g \varepsilon_r \tag{8}$$

in which the dot denotes the rating function.

In the heating process, the stored strain is given to be released gradually in proportion to the volume fraction of the glassy phase with respect to the preceding glassy phase. The strain storage is expressed as:

$$\dot{\varepsilon}_{in} = \frac{\dot{\xi}_g}{\xi_g} \varepsilon_{in} \tag{9}$$

To derive the stress state, the second law of thermo-dynamics in the sense of the Clausius–Duhem inequality should be satisfied. In this model, ε and T are selected as external control variables, while $\varepsilon_g, \varepsilon_r, \varepsilon_{in}$, and ξ_g are internal variables. Considering Helmholtz free energy density functions, stress σ can be derived as:

$$\sigma = \sigma_g = \sigma_r \tag{10}$$

where

$$\sigma_g = C_g \varepsilon_g, \quad \sigma_r = C_r \varepsilon_r \tag{11}$$

where C is the elasticity matrix defined as:

$$C = \frac{E}{(1+v)(1-2v)} \begin{bmatrix} 1-v & v & v & 0 & 0 & 0 \\ v & 1-v & v & 0 & 0 & 0 \\ v & v & 1-v & 0 & 0 & 0 \\ 0 & 0 & 0 & \frac{(1-2v)}{2} & 0 & 0 \\ 0 & 0 & 0 & 0 & \frac{(1-2v)}{2} & 0 \\ 0 & 0 & 0 & 0 & 0 & \frac{(1-2v)}{2} \end{bmatrix} \tag{12}$$

By substituting Equation (11) in Equation (5), we obtain the stress as:

$$\sigma = C_e(\varepsilon - \varepsilon_{in} - \varepsilon_{th}) \tag{13}$$

where C_e is the equivalent elasticity tensor and is expressed as:

$$C_e = (S_r + \xi_g(S_g - S_r))^{-1} \tag{14}$$

in which S denotes the inverse matrix of $C(S = C^{-1})$, so-called the "compliance matrix".

The non-linear SMP behavior can be treated as an explicit time-discrete stress/strain-temperature driven problem. The time domain $[0, t]$ is divided into subdomains, and the equation is solved in the local domain $[t^k, t^{k+1}]$. The superscript $k+1$ for all variables denotes the current step, while the superscript k indicates the previous step. The inelastic strain can be calculated by applying the linearized implicit backward Euler integration method to the flow rule. Thus, Equations (8) and (9) can be discretized as:

$$\varepsilon_{in}^{k+1} = \varepsilon_{in}^k + \Delta \xi_g^{k+1} \varepsilon_r^{k+1} \tag{15}$$

$$\varepsilon_{in}^{k+1} = \varepsilon_{in}^k + \frac{\Delta \xi_g^{k+1}}{\xi_g^{k+1}} \varepsilon_{in}^{k+1} \tag{16}$$

where

$$\Delta \xi_g^{k+1} = \xi_g^{k+1} - \xi_g^k \tag{17}$$

By substituting Equations (11) and (13) into Equations (15) and (16) along with a mathematical simplification, we can explicitly update the inelastic strain for the cooling and heating processes. For the cooling process, we can write:

for stress control:

$$\varepsilon_{in}^{k+1} = \varepsilon_{in}^k + \Delta \xi_g^{k+1} S_r^{k+1} \sigma^{k+1} \tag{18}$$

for strain control:

$$\varepsilon_{in}^{k+1} = (I + \Delta \xi_g^{k+1} S_r^{k+1} C_e^{k+1})^{-1} (\varepsilon_{in}^k + \Delta \xi_g^{k+1} S_r^{k+1} C_e^{k+1} (\varepsilon^{k+1} - \varepsilon_{th}^{k+1})) \tag{19}$$

For the heating process, Equation (16) can be simplified as:

$$\varepsilon_{in}^{k+1} = \frac{\xi_g^{k+1}}{\xi_g^k} \varepsilon_{in}^k \tag{20}$$

Now, by substituting the updated inelastic strain into Equation (15), the stress–strain relationship for heating and cooling processes can be obtained as:

$$\sigma^{K+1} = C_D^{k+1}(\varepsilon^{k+1} - \delta \varepsilon_{in}^{k+1} - \varepsilon_{th}^{k+1}) \tag{21}$$

where elasticity tensor C_D and the δ parameter for the heating and cooling processes are defined as:

$$C_D^{k+1} = (I + \Delta \xi_g^{k+1} S_r^{k+1} C_e^{k+1})^{-1} C_e^{k+1}, \delta = 1 \dot{T} < 0$$
$$C_D^{k+1} = C_e, \delta = \frac{\xi_g^{k+1}}{\xi_g^k} \dot{T} > 0 \qquad (22)$$

3.2. FE Methodology

3.2.1. In-house FE method

A Ritz-based FE solution is implemented to predict the thermo-mechanical behaviors of FG 4D-printed structures. A 3D twenty-node quadratic serendipity hexahedron element is applied to this problem. It has twenty nodes so that eight corner nodes are augmented with twelve side nodes located at the side center. The element also has three degrees of freedom per node ($u_i (i = 1, 2, 3)$. For more details on the FE modeling, one may refer to [23].

3.2.2. FE Abaqus

The results of the DMA test in the form of a temperature-dependent modulus are introduced in the FE Abaqus software. Table 2 shows the value of Young's modulus at different temperatures.

Table 2. Young's modulus for different temperatures implemented in the finite element (FE) Abaqus.

T (°C)	30	40	50	60	70	80	90
E (MPa)	3350	3280	3166	2554	48	18	14

To model the straight beam-like samples with FG features, they are discretized to five sections where each section has a different thermal expansion coefficient. Figure 6 shows a straight beam-like sample in a discretized form with different thermal expansion coefficients.

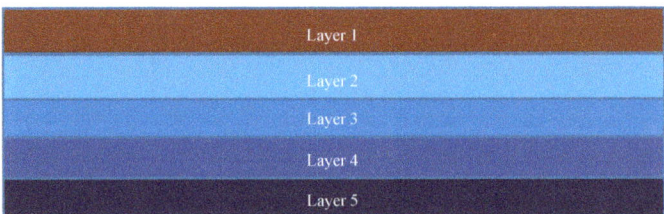

Figure 6. Discretized straight beam-like sample.

In Figure 6, layer 1 indicates the last printed layer, while layer 5 is the first layer that has been 3D-printed.

4. Results and Discussion

In this section, the experimental and numerical analyses of the self-bending/morphing structures are presented. The samples are heated by dipping into the hot water at a prescribed temperature of 85 °C that is greater than the transition temperature by 22 °C. Three straight beam-like samples with dimensions of (30 × 1.6 × 1) mm are printed at liquefier temperature $T_{in} = 190$ °C and the print speed of $S_p = 20\,mm/s$, $S_p = 40\,mm/s$ and $S_p = 70\,mm/s$, respectively. The configuration of the three samples after the heating–cooling process is depicted in Figure 7. As it can be seen, the samples self-bend when dipping in the hot water. The observed self-bending is due to an unbalanced pre-strain induced during 3D printing and deposited through the thickness direction. Unbalancing in the through-the-thickness pre-strain distribution leads to a mismatch in the free-strain recovery, producing curvatures and revealing a transformation from a temporary straight shape to a permanent curved shape. It should be mentioned that curved beams could be manually programmed again to get another temporary shape and reveal shape memory effects upon heating. It is also found that the pre-strain has an increasing trend through the thickness from the lower to the upper layer that leads the beam to be changed upward. This self-bending is such that the samples with higher printing speeds have larger bending angles. One of the reasons that can explain this trend is that more speed provides more mechanical loading that may induce more pre-strain. In fact, the FDM printing process shows the capability of both fabrication and hot programming at the same time. To characterize the deformed shape, we define three geometric parameters (R_1, R_2, R_3) that can describe the deformed shapes. R_3, R_2 and R_1 denote the outer length, the opening, and the depth of the mid-surface of the deformed sample, respectively.

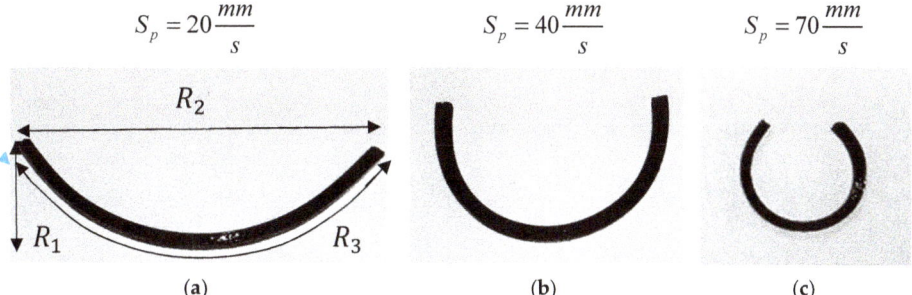

Figure 7. Deformed configurations of the beams printed with different speeds after the heating–cooling process.

Then, the FE Abaqus software is used to model the printed samples. For this purpose, the element type C3D8T is used, and the sample is discretized into five sections with different thermal expansion coefficients. The thermal expansion coefficient on each layer is chosen to obtain the deformed configuration for the specific printing speed. Table 3 shows the thermal expansion coefficient of each layer for different printing speeds. Figure 8 also shows the results from the FE Abaqus simulation.

Table 3. Thermal expansion coefficients for different printing speeds.

$\alpha_i (1/°C)$	$S_p (mm/s)$		
	20	40	70
α_1	−0.00400	−0.00480	−0.00680
α_2	−0.00320	−0.00384	−0.00544
α_3	−0.00240	−0.00288	−0.00408
α_4	−0.00160	−0.00192	−0.00272
α_5	−0.00080	−0.00096	−0.00136

Table 4 also lists the geometric parameters obtained from the experiments, FE Abaqus, and in-house FE code. It is found that simulation results of Abaqus are in a good agreement with the characteristics observed in the experiments and the in-house FE solution. It validates the reliability of the SMP programming by considering FG thermal expansion in the FE Abaqus.

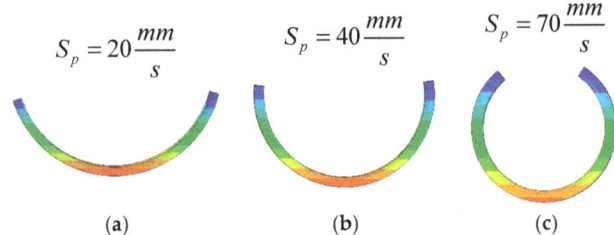

Figure 8. The FE Abaqus simulation of the self-bending beams illustrated in Figure 7.

Table 4. Geometric parameters of the beams printed with different speeds after the heating–cooling process.

Method	V_p (mm/s)	R_1 (mm)	R_2 (mm)	R_3 (mm)
Experiments	20	8.20	21.00	29.50
	40	9.60	12.40	29.10
	70	10.70	5.90	28.50
FE Abaqus	20	8.31	21.20	29.55
	40	9.75	12.34	29.00
	70	10.57	6.08	28.60
In-house FE method	20	8.32	21.30	29.47
	40	9.79	12.41	29.01
	70	10.66	6.02	28.49

Next, potential applications of self-bending primitives are demonstrated. First, we design a flat sheet with dimensions of (50 × 30) mm, reinforced by three straight beams that are printed with different speeds. Figure 9 shows the experimental results of the deformed configuration of the structure after heating up to 65 °C and then cooling down to room temperature. Young's modulus of the PLA at 85 °C is very low. Therefore, if the structure is heated up to 85 °C, the beams become very soft and the paper sheet under tension returns the beams to the undeformed configuration. That is why the structure should be heated up to 65 °C. As observed in Figure 9, the bending angles are different for the beams printed with different speeds. Due to this fact, the sheet is bent along the central line with different angles and deformed into a conical panel by heating. This can be considered as a demonstration of a 2D-to-3D shape-shifting by the self-morphing mechanism.

Figure 9. Configuration of the flat sheet reinforced by three straight 4D-printed beams after the heating–cooling process.

The composite structure, including the main flat sheet reinforced by three straight beams with different pre-strain levels, is modeled by the FE Abaqus. The interaction between the beams and the paper sheet is of the Tie type, which is a perfect bond between the beams and the paper sheet. After determining the thermal boundary conditions similar to the experimental conditions, the structure is heated up to 65 °C and then cooled down to the room temperature. Figure 10 represents simulation results of the deformed configuration that properly match with the experimental shape. As expected, the beam that is 4D printed with a lower speed has a lower bending angle, while the beams printed faster produce greater bending angles. Similar to the experimental results, the paper sheet bends and transforms into a conical panel with the self-morphing feature.

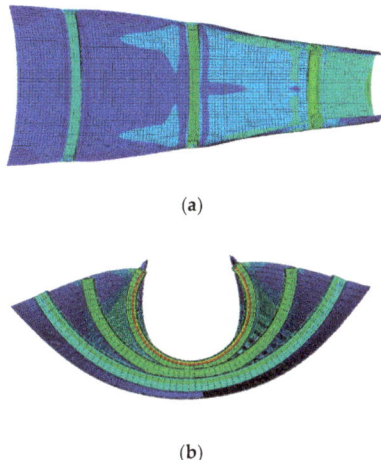

(a)

(b)

Figure 10. The FE Abaqus simulation of the self-morphing composite structure illustrated in Figure 9.

For the second example, we consider a plus-like structure printed at $S_p = 40\ mm/s$. This structure consists of two perpendicular beams with dimensions of (30 × 1.6 × 1) mm. The structure is heated up to 85 °C and then cooled down to room temperature. Figure 11 illustrates the experimental results of the deformed configuration after the heating–cooling process. It can be found that this element has the potential to be used as a flexible self-bending gripper for future mechanical/biomedical devices fabricated by the 4D printing technology. The bending of the gripper can be controlled by changing the printing speed. For example, if a sample is printed at $S_p = 70\ mm/s$, the bending of the gripper becomes greater. This means that the printing speed can be manipulated to get a desired angle. To model this structure with the FE Abaqus, it is divided into five sections through its thickness, and the thermal boundary conditions are chosen similar to the experimental conditions. The plus-like structure is heated up to 85 °C and then cooled down to room temperature. Figure 12 shows the deformed configuration obtained from the simulation.

Figure 11. Deformed configuration of the flat plus-like structure after the heating–cooling process.

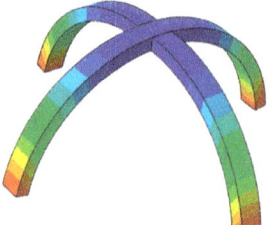

Figure 12. The FE Abaqus simulation of the self-bending gripper illustrated in Figure 11.

The comparison studies in Figures 7–12 revealed the high accuracy of the 3D FE method in Abaqus in replicating the experimental observations. In the following studies, this digital Abaqus tool is implemented to simulate various self-bending devices.

Figure 13 shows a flower-like structure composed of a flat paper sheet and eight straight beams. The dimensions of the beams are (30 × 1.6 × 1) mm. The printed beams are at a 10 mm distance from the center of the structure. The beam structures are printed on the paper such that the first printed layer is directly connected to the paper. To model the interactions between the beams and the paper sheet, a Tie-type interaction is assumed. The beam-like structures are 4D printed with different speeds for three different case studies. The configuration of the flower-shaped structure reinforced with beams printed with different speeds after the heating–cooling process is displayed in Figure 14. As it can be seen, when the structure is heated up to 65 °C, it is bent towards the interior layer and the structure transforms into a flower shape.

Figure 13. Undeformed configuration of a flower-shaped structure.

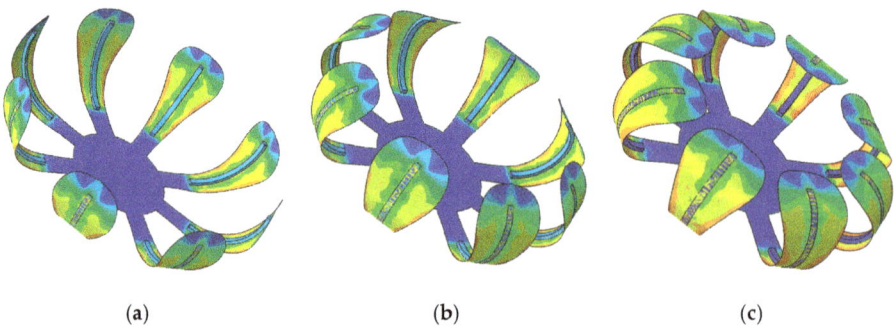

Figure 14. FE Abaqus simulation of the self-morphing flower-shaped structure after the heating–cooling process: (**a**) $S_p = 40\,mm/s$; (**b**) $S_p = 70\,mm/s$; (**c**) $S_p = 70\,mm/s$ (the edge shown on the outer layer of the paper sheet demonstrates the sample place on the interior layer).

The results presented in Figure 14a show that the deformed configuration for the case of $S_p = 20\,mm/s$ has a lower bending angle. As it can be observed, by increasing the printing speed, the flower further closes. As another example, a bunch of beams with dimensions of $(30 \times 1.6 \times 1)$ mm are diagonally printed over a rectangular paper sheet with dimensions of (230×21) mm, as shown in Figure 15. The angle between the paper sheet and beams is 45°. The beams are connected to the paper such that the first printed layer is directly connected to the paper. The interaction between the paper sheet and beams is of the Tie type. This structure is heated up to 65 °C and then cooled down to the room temperature. The beam-like structures are 4D printed with different speeds for three different case studies. Figure 16 illustrates the configuration of the rectangular paper sheet reinforced with the beams fabricated with different speeds after the heating–cooling process.

Figure 15. Configuration of the rectangular paper sheet with patterned oblique beams.

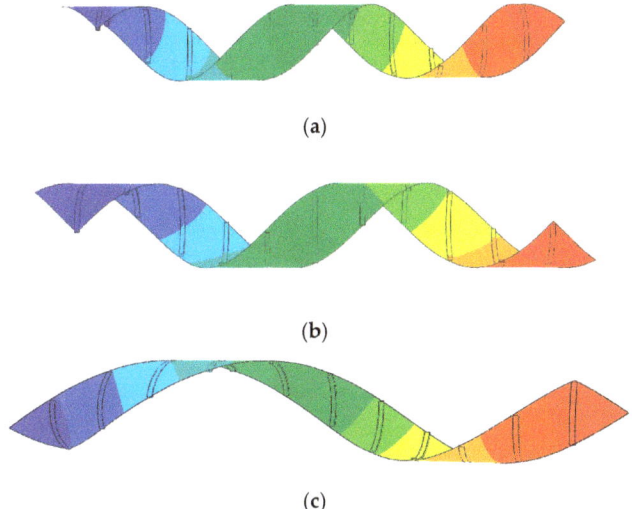

Figure 16. The FE Abaqus simulation of the self-rolling helix after the heating–cooling process: (a) $S_p = 20\,mm/s$, (b) $S_p = 40\,mm/s$, (c) $S_p = 70\,mm/s$ (the edge shown on the outer layer of the paper sheet demonstrates the sample place on the interior layer).

Figure 16 reveals that the structure, initially in a flat state, transforms into a helix upon heating, revealing a self-rolling feature. It is observed that enhancing the printing speed increases the pitch. Therefore, by changing the printing speed, the pitch can be controlled. Moreover, by changing the angle between the paper sheet and the 4D-printed beams, the geometry of the self-rolling helix could be changed.

5. Conclusions

The aim of this paper was to develop self-bending/morphing/rolling structures fabricated by FG 4D printing and introduce a novel simple computational tool for replicating their thermo-mechanical behaviors. The concept was based on the understanding of SMP thermo-mechanics and programming the material via common FDM 3D printing technology. Structural primitives with self-bending 1D-to-2D features were 4D printed and then employed as actuation elements to engineer complex structures with 2D-to-3D shape-shifting via self-bending/morphing/rolling mechanisms. The influences of the printing speed on the self-morphing characteristics were investigated in detail. 1D-to-2D and 2D-to-3D shape transformations were simulated by introducing a straightforward method into the commercial FE software package of Abaqus that is much simpler than writing a UMAT subroutine or an in-house FE solution. The 4D-printed materials were modeled as FG materials whose thermal expansions varied through the thickness direction. The accuracy of the proposed approach was verified by a comparison study with experiments and results obtained from the in-house FE solution. Due to the absence of a similar concept, numerical approach, or results in the specialized literature, this paper is likely to pave the way for designing self-bending/morphing/rolling adaptive structures by 4D FDM printing technology.

Author Contributions: Conceptualization, M.B.; data curation, S.N.; funding acquisition, M.B.; investigation, M.B. and R.N.; methodology, M.B. and R.N.; project administration, M.B.; resources, M.B. and R.N.; supervision, M.B., A.Z., and M.F.; validation, M.B., R.N., A.Z., and M.F.; writing—original draft, M.B. and R.N.; writing—review & editing, A.Z., M.F., and S.N.

Funding: This research received no external funding.

Conflicts of Interest: The authors declare no conflict of interest.

References

1. Rengier, F.; Mehndiratta, A.; Von Tengg-Kobligk, H.; Zechmann, C.M.; Unterhinninghofen, R.; Kauczor, H.U.; Giesel, F.L. 3D printing based on imaging data: review of medical applications. *Int. J. Comput. Assist. Radiol. Surg.* **2010**, *5*, 335–341. [CrossRef] [PubMed]
2. Hull, C.W. Apparatus for Production of Three-Dimensional Objects by Stereolithography. U.S. Patent 4,575,330, 8 August 1984.
3. Tymrak, B.; Kreiger, M.; Pearce, J.M. Mechanical properties of components fabricated with open-source 3-D printers under realistic environmental conditions. *Mater. Des.* **2014**, *58*, 242–246. [CrossRef]
4. Sun, Q.; Rizvi, G.; Bellehumeur, C.; Gu, P. Effect of processing conditions on the bonding quality of FDM polymer filaments. *Rapid Prototyp. J.* **2008**, *14*, 72–80. [CrossRef]
5. Tran, P.; Ngo, T.D.; Ghazlan, A.; Hui, D. Bimaterial 3D printing and numerical analysis of bio-inspired composite structures under in-plane and transverse loadings. *Compos. Part B-Eng.* **2017**, *108*, 210–223. [CrossRef]
6. Melnikova, R.; Ehrmann, A.; Finsterbusch, K. 3D printing of textile-based structures by Fused Deposition Modelling (FDM) with different polymer materials. *IOP Conf. Ser. Mater. Sci. Eng.* **2014**, *62*, 1–6. [CrossRef]
7. Caulfield, B.; McHugh, P.; Lohfeld, S. Dependence of mechanical properties of polyamide components on build parameters in the SLS process. *J. Mater. Process. Technol.* **2007**, *182*, 477–488. [CrossRef]
8. Garcia, C.R.; Correa, J.; Espalin, D.; Barton, J.H.; Rumpf, R.C.; Wicker, R.; Gonzalez, V. 3D printing of anisotropic metamaterials. *Prog. Electromagn. Res.* **2012**, *34*, 75–82. [CrossRef]
9. Endo, F.; Saiga, N. Thermoplastic Polymer Composition and Medical Devices Made of the Same. E.U. Patent 0568451, 28 April 1995.
10. Wang, Q.; Tian, X.; Huang, L.; Li, D.; Malakhov, A.V.; Polilov, A.N. Programmable morphing composites with embedded continuous fibers by 4D printing. *Mater. Des.* **2018**, *155*, 404–413. [CrossRef]
11. Hu, J.; Meng, H.; Li, G.; Ibekwe, S.I. A review of stimuli-responsive polymers for smart textile applications. *Smart Mater. Struct.* **2012**, *21*, 053001. [CrossRef]
12. Zadpoor, A.A. Mechanical performance of additively manufactured meta-biomaterials. *Acta Biomater.* **2018**, *85*, 41–59. [CrossRef]
13. Lakes, R. Foam structures with a negative Poisson's ratio. *Science* **1987**, *235*, 1038–1041. [CrossRef]
14. Surjadi, J.U.; Gao, L.; Du, H.; Li, X.; Xiong, X.; Fang, N.X.; Lu, Y. Mechanical Metamaterials and Their Engineering Applications. *Adv. Eng. Mater.* **2019**, *20*, 1800864. [CrossRef]
15. Wang, K.; Chang, Y.-H.; Chen, Y.; Zhang, C.; Wang, B. Designable dual-material auxetic metamaterials using three-dimensional printing. *Mater. Des.* **2015**, *67*, 159–164. [CrossRef]
16. Mirzaali, M.; Caracciolo, A.; Pahlavani, H.; Janbaz, S.; Vergani, L.; Zadpoor, A. Multi-material 3D printed mechanical metamaterials: Rational design of elastic properties through spatial distribution of hard and soft phases. *Appl. Phys. Lett.* **2018**, *113*, 241903. [CrossRef]

17. Yang, H.; Wang, B.; Ma, L. Designing hierarchical metamaterials by topology analysis with tailored Poisson's ratio and Young's modulus. *Compos. Struct.* **2019**, *214*, 359–378. [CrossRef]
18. Bodaghi, M.; Damanpack, A.; Hu, G.; Liao, W.H. Large deformations of soft metamaterials fabricated by 3D printing. *Mater. Des.* **2017**, *131*, 81–91. [CrossRef]
19. Lee, A.Y.; An, J.; Chua, C.K. Two-way 4D printing: a review on the reversibility of 3D-printed shape memory materials. *Engineering* **2017**, *3*, 663–674. [CrossRef]
20. Tibbits, S.; McKnelly, C.; Olguin, C.; Dikovsky, D.; Hirsch, S. 4D Printing and universal transformation. In Proceedings of the 34th Annual Conference of the Association for Computer Aided Design in Architecture, Los Angeles, CA, USA, 23–25 October 2014; pp. 539–548.
21. Zhang, Q.; Zhang, K.; Hu, G. Smart three-dimensional lightweight structure triggered from a thin composite sheet via 3D printing technique. *Sci. Rep.* **2016**, *6*, 22431. [CrossRef]
22. Jamal, M.; Kadam, S.S.; Xiao, R.; Jivan, F.; Onn, T.M.; Fernandes, R.; Nguyen, T.D.; Gracias, D.H. Bio-origami hydrogel scaffolds composed of photocrosslinked PEG bilayers. *Adv. Healthc. Mater.* **2013**, *2*, 1142–1150. [CrossRef]
23. Bodaghi, M.; Damanpack, A.; Liao, W.H. Adaptive metamaterials by functionally graded 4D printing. *Mater. Des.* **2017**, *135*, 26–36. [CrossRef]
24. Teoh, J.; An, J.; Feng, X.; Zhao, Y.; Chua, C.; Liu, Y. Design and 4D printing of cross-folded origami structures: A preliminary investigation. *Materials* **2018**, *11*, 376. [CrossRef]
25. Yang, C.; Boorugu, M.; Dopp, A.; Ren, J.; Martin, R.; Han, D.; Choi, W.; Lee, H. 4D printing reconfigurable, deployable and mechanically tunable metamaterials. *Mater. Horiz.* **2019**. [CrossRef]
26. Bodaghi, M.; Liao, W.H. 4D printed tunable mechanical metamaterials with shape memory operations. *Smart Mater. Struct.* **2019**, *28*, 045019. [CrossRef]
27. Zolfagharian, A.; Kaynak, A.; Yang Khoo, S.; Zhang, J.; Nahavandi, S.; Kouzani, A. control-oriented modelling of a 3D-printed soft actuator. *Materials* **2019**, *12*, 71. [CrossRef]
28. Bertoldi, K.; Vitelli, V.; Christensen, J.; van Hecke, M. Flexible mechanical metamaterials. *Nat. Rev. Mater.* **2017**, *2*, 1–11. [CrossRef]
29. Yu, K.; Yin, W.; Sun, S.; Liu, Y.; Leng, J. Design and analysis of morphing wing based on SMP composite. *Proc. SPIE* **2009**, *7290*, 72900S.
30. Tao, R.; Yang, Q.S.; He, X.Q.; Liew, K.M. Parametric analysis and temperature effect of deployable hinged shells using shape memory polymers. *Smart Mater. Struct.* **2016**, *25*, 115034. [CrossRef]

© 2019 by the authors. Licensee MDPI, Basel, Switzerland. This article is an open access article distributed under the terms and conditions of the Creative Commons Attribution (CC BY) license (http://creativecommons.org/licenses/by/4.0/).

MDPI
St. Alban-Anlage 66
4052 Basel
Switzerland
Tel. +41 61 683 77 34
Fax +41 61 302 89 18
www.mdpi.com

Materials Editorial Office
E-mail: materials@mdpi.com
www.mdpi.com/journal/materials

www.ingramcontent.com/pod-product-compliance
Lightning Source LLC
LaVergne TN
LVHW072000080526
838202LV00064B/6806